Engineering Graphics and Descriptive Geometry in 3-D

G. F. PEARCE

Science and Technology Division
CAMOSUN COLLEGE
1950 LANSDOWNE ROAD
VICTORIA, B.C. V8P 5J2

The Macmillan Company of Canada

©1977, G. F. Pearce

All rights reserved—no part of this book may be reproduced in any form without permission from the publisher, except by a reviewer who wishes to quote brief passages in connection with a review written for inclusion in a magazine or newspaper.

Canadian Cataloguing in Publication Data

Pearce, Gordon F., 1917 –
    Engineering graphics and descriptive geometry in 3-D

Bibliography: p.
Includes Index.
ISBN 0-7705-1536-3

1. Engineering graphics. 2 Geometry, Descriptive.
I. Title

T353.P43   604.2   C77-001120-9

Printed in Canada for
The Macmillan Company of Canada Limited

Contents

Preface i

Analglyph Drawings i

1. **Three-Dimensional Solids** 1
   Geometrical Solids, Engineering Solids, Plane Geometrical Constructions

2. **Fundamentals of Projection** 13
   Terminology, Projection of an Object, Orthographic Projection

3. **Multiview Orthographic Projection** 17
   Unfolding of Projection Planes, Space Quadrants, First- and Third-Angle Projection

4. **Points and Lines in Three-Dimensional Space** 35
   Location of a Point, Straight Lines

5. **Auxiliary Planes** 49
   True Length, Bearing, Slope, Inclined or Double Auxiliary Planes

6. **Perpendicular Lines** 57
   Intersecting and Non-intersecting

7. **Parallel Lines** 65
   Parallel Projections

8. **Plane Surfaces** 67
   Edge View, True Slope, True Shape, Dihedral Angle, Angle between Line and Plane, Line Perpendicular to Plane, Visibility

9. **Distances Between Points, Lines, and Planes** 89
   Line and Plane Methods, Horizontal Distance, Point to a Plane

10. **Piercing Points** 103
    Edge-View and Cutting-Plane Methods

11. **Rotation** 111
    Point about a Line, Line about a Line, Angle between Line and Plane, Line at Specified Angles, True Shape, Dihedral Angle

12. **Pictorial Views** 129
    Oblique, Isometric, and Perspective Projection

13. **Vector Geometry – Coplanar Vectors** 163
    Addition, Resolution, Concurrent Vectors, Bow's Notation, Beams, Trusses, Method of Joints, Method of Sections

14. **Vector Geometry – Non-coplanar Vectors** 189
    Addition, Resultant, Equilibrium, Trial Solution, Edge-View Method

15. **Kinematics** 209
    Displacement, Position Synthesis, Relative Velocity, Space Mechanisms

16. **Graphical Calculus** 227
    Differentiation, Integration

17. **Short-Range Photogrammetry** 239
    The Photographic View, Measurements from Photographs

18. **Computers, Calculators, and Graphics** 257
    General Procedure, Isometric and Perspective Projections

19. **Intersections and Developments** 277
    Edge-View and Cutting-Plane Methods for Intersections, Developments, Cylinder, Cone, Elbow Joints, Triangulation, Warped Surfaces, Convolutes, Gore and Zone Development of a Sphere

20. **Interpreting Engineering Drawings** 299
    Standard Views, Linework, Sectional Views, Dimensioning, Surface Finish, Screw Threads, Welding, Piping, Electrical, Structural, Architectural Drawings, Simplified Drawings, Assembly, Limits and Fits, Geometrical Tolerances, Applications of Geometrical Tolerances

21. **Freehand Technical Sketching** 339
    Pencil and Line Techniques, Crating, Proportion, Pictorial, Oblique, Isometric and Perspective, Exploded Pictorial, Shading

Appendix **A – Equipment** 355

Appendix **B – Accuracy** 361

Appendix **C – Lettering** 365

Bibliography 370

Index 371

Preface

(i) Introduction

This book has been developed as an aid to understanding engineering drawings; to introduce students to the skills required to make drawings and freehand sketches; and to show the use of descriptive geometry to solve engineering problems involving three-dimensional space.

The emphasis is on the comprehension of engineering drawings rather than on drafting techniques. This is because many more people are involved in reading and interpreting engineering drawings than in actually producing them.

Descriptive geometry may be considered to be the use of graphical methods to solve problems involving three-dimensional space. The type of problem studied involves the spatial relationship between points, lines, planes, and solid objects.

Chapter 20, Interpreting Engineering Drawings, may be studied at an early stage, perhaps even following Chapter 3. Chapter 21, Freehand Technical Sketching, should be taught from the beginning of the course, concurrently with the early material on projection. Chapter 18, Computers, Calculators, and Graphics, serves as an introduction to computer graphics. A knowledge of isometric drawing and perspective projection is necessary before this material can be considered.

While most of the objects which concern engineers and architects are solid, that is three-dimensional, they must be represented on plane, that is two-dimensional flat drawings. This at once produces a difficulty in explaining the drawing techniques needed to show solid figures on flat paper. Teachers have alleviated this problem by using actual models to aid in demonstrating the geometrical relationships and constructions required to solve problems in descriptive geometry and graphical representation. In these models, lines are indicated by thin rods, threads, or wires, and surfaces by glass or plastic sheets. This book uses stereoscopic pairs of drawings called *analglyphs* which provide the student with visual models which are available any time and in a greater variety and detail than is possible in most physical models used in classrooms. The student can refer to the models anywhere that he has the book. The stereoscopic pairs of drawings are viewed through coloured filters to create the three-dimensional visual models in a manner to be explained in the next section.

S.I. units are used throughout the text.

(ii) Analglyph Drawings and the Use of Filter Spectacles

Both normal black and white drawings and analglyph drawings in red and green on white will be used throughout this text. The analglyphs may be considered as spatial models and should be referred to as directed in the text as they are used as visual aids to explain the material.

An analglyph drawing combines the two images of an object as seen by the right and left eye. When viewed through the red and green filter "spectacles", found in the back of the book, the analglyph figure permits the object to be recreated in space above the surface of the page (see Figure 1). This is because the right eye will see only the image designed for that eye as the red filter permits only the green image to be seen. Similarly the green filter provides the left eye with its appropriate image. The combination of the two images produces a stereoscopic effect thus recreating the model on and just above the surface of the page. Because of changes in the source and direction of the light,

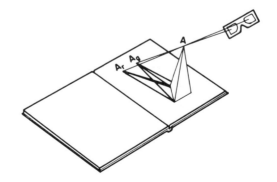

Figure 1 Formation of Three-Dimensional Analglyph Figure

sometimes one image is not completely eliminated; in this case a faint ghost image is presented. It can be ignored by concentrating the attention on the darker lines. The best position for the light to fall on the page is usually from behind or from one side of the viewer. The spectacles should be held in the right hand by the holding-tab portion of the spectacle so that the right eye looks through the red filter. The red filter is nearest the holding-tab. Avoid touching the filters as they are difficult to clean and may become scratched. A person who is colour blind will, in most cases, be able to see the three-dimensional figures as readily as a person who is not colour blind.

The stereoscopic model will appear to change shape as the head is moved. The models have been drawn to be viewed at an angle of about 50° from a comfortable reading distance with the book flat on the desk. The correct lateral position can be determined by shifting the head or book until a vertical line is straight up and down and not slanted. Nearly every analglyph drawing has an obvious vertical line which may be used as a guide to find the proper viewing position.

The analglyph drawings have been carefully constructed to scale. In this respect any line may be measured. However, in order to measure lines and angles the correct vertical viewpoint must be found as moving the head up or down changes the lengths of lines. A transparent protractor and a transparent ruler are convenient aids. The ruler must be held parallel to the line being measured. As an example the geometric solids in Figure 1.3 are all 5 cm high when viewed from the correct position. It is usually not necessary to measure the lines or angles except as a matter of interest because the geometric relations depicted in any of the models are readily apparent from many viewing positions. It may take several seconds before the figures appear to stand up out of the page the first few times they are viewed through the spectacles. After a very few sessions the three-dimensional models appear immediately upon viewing through the filters.

Science and Technology Division
CAMOSUN COLLEGE
1950 LANSDOWNE ROAD
VICTORIA, B.C. V8P 5J2

1. Three-Dimensional Solids

1.1 GEOMETRICAL SOLIDS

Two dimensions only are required to specify a point on a plane geometric figure. Some examples are triangles, squares, circles, and hexagons as shown in Figure 1.1. The point A for example is located from the origin by the rectangular coordinates x_A and y_A.

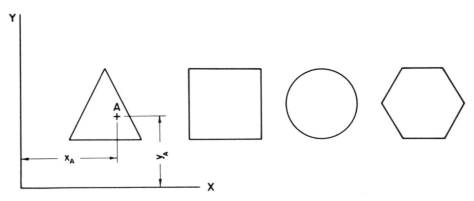

Figure 1.1 Plane Geometrical Figures

Solid objects however require three dimensions to specify any point either on or inside of them. Some regular solids are shown in the analglyph Figure 1.2. These particular geometric solids are a cube, a triangular prism, and a right prism or rectangular parallelepiped. The point B on the cube is located from the origin O by the rectangular coordinates x_B, y_B, and z_B.

These figures are accurately drawn so that when they are viewed from the correct position the cube will be 25 mm in each direction, the others will have 25 mm sides and will appear 57 mm high. If you find that the figure is higher than 57 mm then your eyes are too high above the page. Similarly if the figure appears less than 57 mm high then your eyes are too low. It is not essential that the exact position be found as the geometrical relationships are apparent from any reasonably normal reading position.

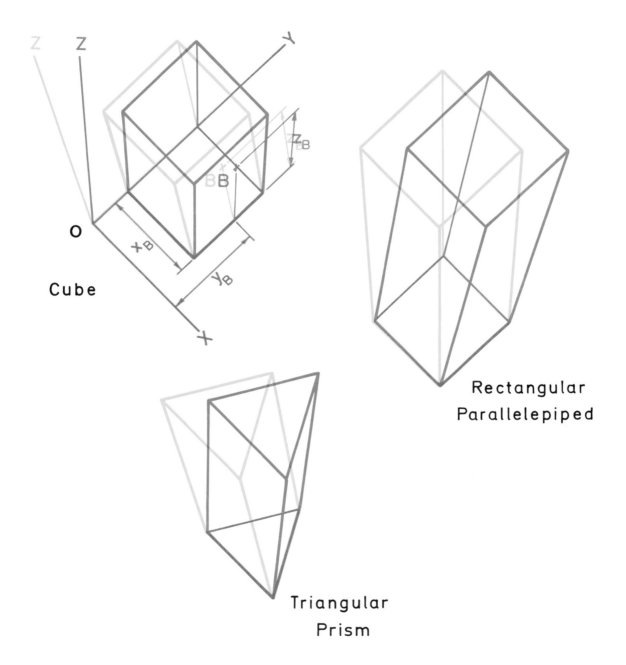

Figure 1.2 Three-Dimensional Geometrical Solids

Some additional geometrical solids are shown in Figure 1.3. These include an upright hexagonal prism, a right square pyramid, a right cone, a right cylinder, and an oblique triangular pyramid, all 5 cm high.

It may be noted that the models appear to move and to change shape as the head is moved. The correct viewing position may be quickly established by moving the head until the lines which should be vertical do take up a vertical position. In addition, for perfect viewing the plane of the page must be flat. In most of the models in this text it is not necessary to find and use the precisely correct viewing position. Any reasonable viewing position with the book flat and in front of you will suffice. This is because the geometrical relationships being explained will be apparent from any normal viewing and reading position.

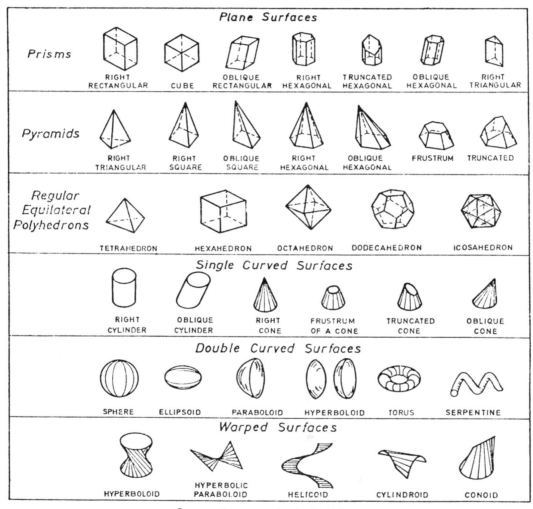

Some Geometrical Solids

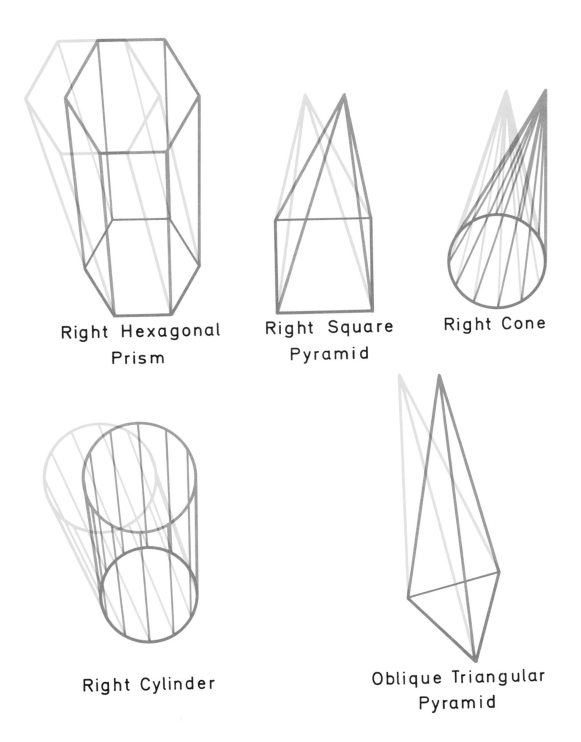

Figure 1.3 Three-Dimensional Geometrical Solids

1.2 ENGINEERING SOLIDS

Solids of engineering interest are not usually made up of purely geometrical shapes but are combinations of solids such as those shown in Figure 1.4 and Figure 1.5.

The small bracket shown full size in Figure 1.4 is shown in three different positions and thus provides us with three different views of it. All the information required to make this part cannot be shown in one view. In order to show all the detail necessary for engineering purposes, a series of views of most parts is required. This book will develop the methods used to represent such three-dimensional objects on two-dimensional paper.

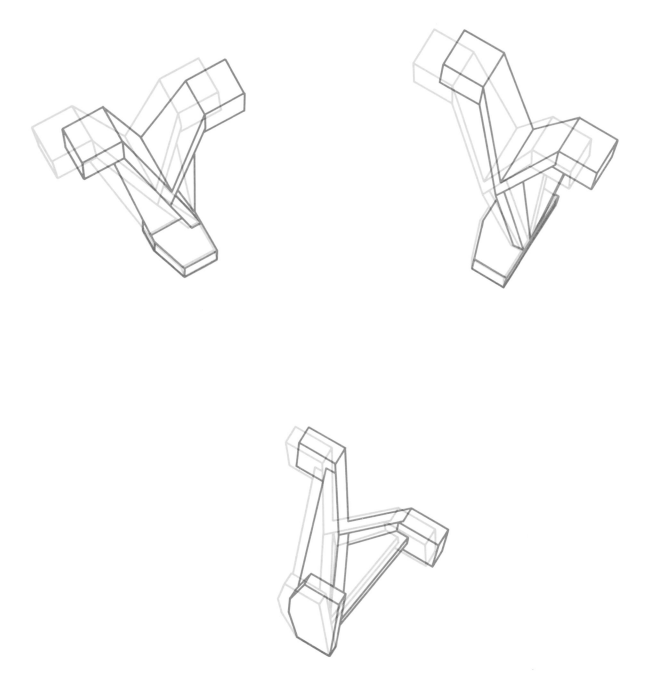

Figure 1.4 Engineering Solid—Mechanical Part

The model in Figure 1.5 is of an artificial satellite or space probe. It is obvious that a large number of drawings would be required to show all the necessary detail needed to manufacture such a complex vehicle. Each part of the object would have to be shown in several different views. In addition, the angles between the members of the supporting frame would have to be determined accurately. Methods of determining these requirements are developed by the use of descriptive geometry.

During the preliminary design stage of the space probe, the forces in the members of the frame would have to be determined. It is necessary to know the magnitude of these forces and also whether each member is in tension or compression. This force analysis is another type of problem which can be solved using the methods of descriptive geometry.

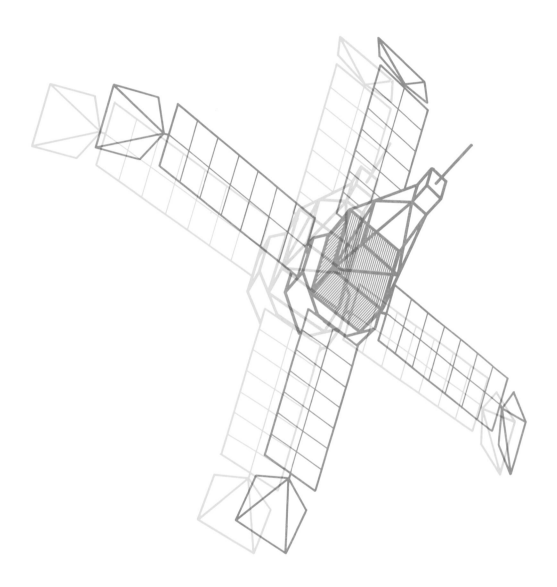

Figure 1.5 Engineering Solid—Space Probe

1.3 PLANE GEOMETRICAL CONSTRUCTIONS

In order to make accurate engineering drawings and to solve problems using descriptive geometry it is often necessary to use various geometrical constructions. Many of these will be familiar to most students, such as using a compass to bisect an angle or to draw a perpendicular to a line. Some of the more frequently used constructions are explained below. In connection with geometrical drawing the student should read the Appendices on Equipment, Accuracy, and Lettering.

To Draw a Line Parallel to a Given Line (Figure 1.6)
The hypotenuse of one drafting triangle (either the 30-60° or the 45°) is placed against the hypotenuse of the other. The two are moved together until one side of the triangle is in exact alignment with the given line. Now hold the lower triangle fixed and slide the other triangle to the required position. Draw the parallel line.

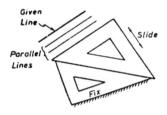

Figure 1.6 Line Parallel to Given Line

To Draw a Line Perpendicular to a Given Line (Figure 1.7)
This method is essentially similar to that described above, see Figure 1.7.

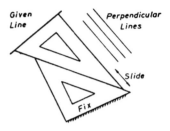

Figure 1.7 Line Perpendicular to Given Line

To Divide a Straight Line into a Number of Equal Parts (Figure 1.8)

Given the line AB, suppose that we wish to divide it into seven equal parts. First draw any line AX of convenient length and making any convenient angle with AB. Using dividers step off seven equal parts on AX. Alternatively a scale may be used and seven equal lengths marked off on AX. Join 7B. Through each point 1, 2, 3, etc. draw lines parallel to 7B to cut the given line in points as shown. These points are equally spaced along AB.

To Draw a Tangent to a Circle from a Point Outside (Figure 1.9)

Given the circle and point P, position a triangle so that one side passes through the point P and is tangent to the circle. Slide the triangle along, against a second triangle, until the other leg passes through the centre of the circle. This leg will intersect the circle at the tangent point.

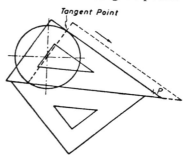

Figure 1.9 Tangent to a Circle

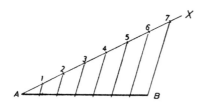

Figure 1.8 Dividing a Line into Equal Parts

To Construct a Regular Polygon in a Circle (Figure 1.10)

Divide a diameter of the circle into the same number of equal parts as the number of sides of the polygon (seven in this example). Construct an equilateral triangle, AB7 on the diameter. Draw the line B2 and let it cut the circle at C. Use the distance AC to step off the vertices of the polygon around the circumference.

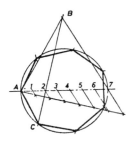

Figure 1.10 Regular Polygon

To Construct a Hexagon within a Circle (Figure 1.11)

Draw a diameter and at each end use the 30-60° triangle to construct chords. Alternatively, use compass or dividers set to the radius of the circle and step off arcs to produce the points of the hexagon.

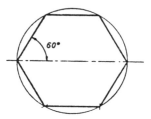

Figure 1.11 Inscribed Hexagon

To Construct a Hexagon given the Distance across the Flats (Figure 1.12)

Draw the inscribed circle. Use the 30-60° triangle to draw tangents to the circle.

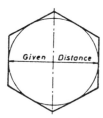

Figure 1.12 Inscribed Circle

Use of the Irregular or French Curve (Figure 1.13)
This device is useful when drawing non-circular lines through a series of plotted points. The line should first be lightly sketched by freehand. Make the line continuous and smooth but do not lose the plotted points. Lay the curve against the line selecting a portion that fits most closely. It is important that the direction of increasing curvature of the French curve is in the same direction as increasing curvature of the line. Now draw a portion of the line stopping short of the point where line and curve appear to coincide. In other words, allow some overlap. Next change the position of the curve to find another section which coincides. Draw another portion. Repeat.

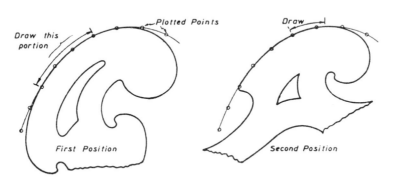

Figure 1.13 Use of Irregular Curves

PROBLEMS
These are mainly exercises to enable students to become familiar with drawing instruments and techniques.

P 1.1 to P 1.3 Construct the figures carefully to the sizes shown.

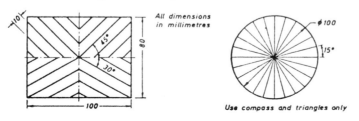

Figure P1.1 *Figure P1.2* *Figure P1.3*

P 1.4 to P 1.6 Lay out the sine curve, cam profile, and wrench as indicated.

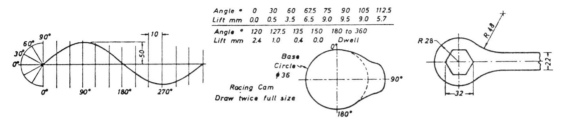

Figure P1.4 *Figure P1.5* *Figure P1.6*

2. Fundamentals of Projection

The basic problem of representing three-dimensional solid objects on a two-dimensional sheet of paper has been solved by means of projecting lines of sight from the object to a plane. The general idea is shown in Figure 2.1. Our flat sheet of drawing paper becomes the plane of projection upon which we represent the object. There are several variations of this general idea. Three basic systems of projection are:
a) Perspective projection
b) Oblique projection
c) Orthographic projection

In order to understand and use each of these systems, the following four elements, illustrated in Figure 2.1, must be kept in mind:
1) Location of the point of sight
2) Direction of the projecting lines
3) Plane of projection
4) Position of the object relative to the plane of projection

2.1 TERMINOLOGY

The *point of sight* is the real or imagined position of the eye of the observer, relative to the object and the plane of projection upon which the particular view is obtained.

The *projecting lines or projectors* are imaginary lines connecting the eye of the observer with points on the object.

The *plane of projection* is the plane surface upon which the object is projected or drawn.

The *object* may be anything real or imagined which it is desired to represent graphically.

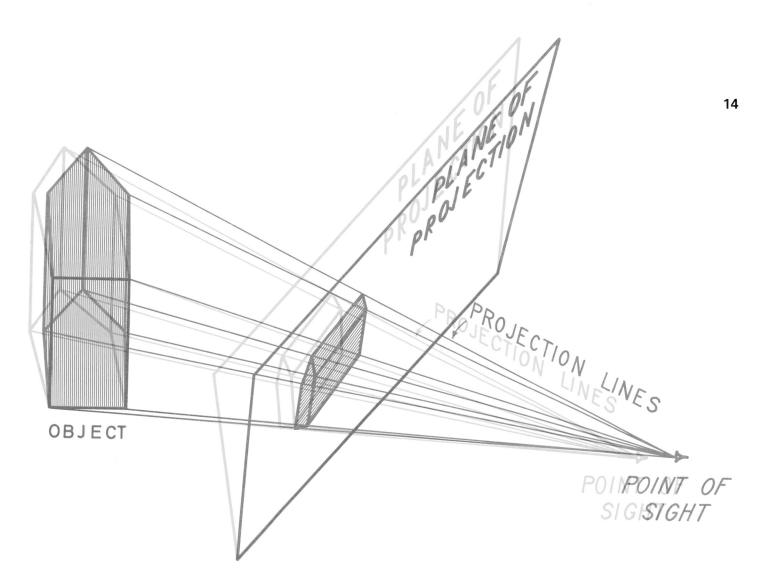

Figure 2.1 Fundamentals of Projection

2.2 PROJECTION OF AN OBJECT

The projection of a point on an object onto the plane of projection is the point at which the line of sight pierces that plane. The projection of the complete object is found by connecting the points where the lines of sight through significant points on the edges of the object pierce the plane of projection.

It is apparent then, that the projection of any object can be obtained by finding the projections of all its significant points, lines, and faces. On the previous page in Figure 2.1 note that the projection of the object appears to be distorted. This is because of the particular position of the object, the projection plane, the lines of sight, and the direction of the projecting lines. The distorted view is generally not suitable for working drawings because measurements of the true size and shape cannot be easily made from such a drawing.

In order to control the distortion, we can alter the four elements which affect the shape of the resulting projected drawing. Two such projection systems which control, but do not eliminate, distortion are: oblique and perspective projection. They will be considered in a later chapter together with isometric projection. A fourth projection system, that used for most working drawings, is called orthographic projection, and will be considered next.

2.3 ORTHOGRAPHIC PROJECTION

In the orthographic projection system, the point of sight is placed at infinity so that the projection lines become parallel. In addition the projection plane is perpendicular to the projection lines. Because of these two conditions, the size of the resulting view of the object on the projection plane does not depend on the distance between the object and the plane. The general arrangement is illustrated in Figure 2.2 in which the object is projected onto a vertical plane. Note that in this view, the true width and height of the object are shown but the depth dimension is not shown. Note also, that the object has been placed so that its major axis, the width, is parallel to the projection plane.

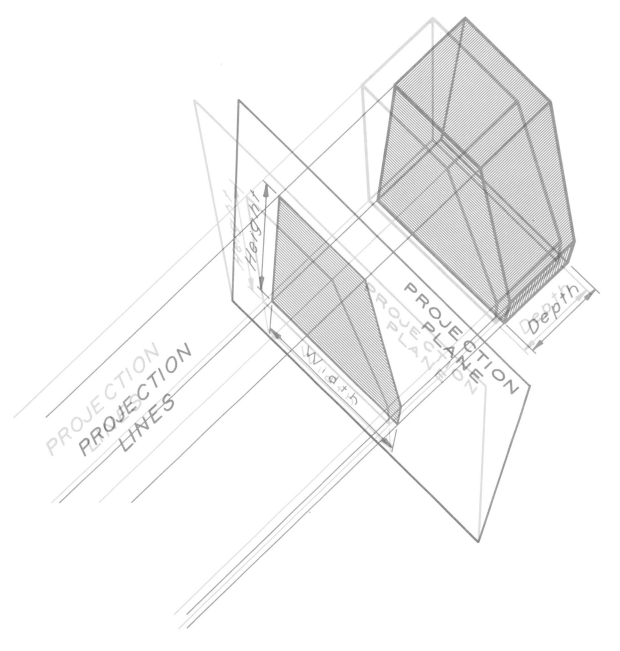

Figure 2.2 Orthographic Projection of a Solid

3. Multiview Orthographic Projection

3.1 GENERAL ARRANGEMENT

A single orthographic view of an object cannot show all the details required to completely illustrate the object. As an example, in Figure 2.2 the depth dimension did not show on the projection plane.

Usually we require more than one projection plane and corresponding views in order to describe an object accurately and completely, so that the size and shape are defined in all the necessary detail so that it can be manufactured.

The position of these additional principal projection planes, and thus of views of an object, is not left to free choice but is determined according to standard systems of drawing. This is so that the person responsible for manufacturing or assembling the object will find the views in the same relation to one another on every drawing. He will not have to ask, for example, "Is this the left end-view or is this the top view?" Adhering to a standard set of rules for technical drawing has been found to improve the efficiency and value of the drawing.

In the standard multiview projection system three mutually orthogonal planes of projection are used. Additional planes (auxiliary planes) may be needed to show all the details and will be discussed later. The object to be portrayed is arranged so that its major axes are parallel or perpendicular to the three orthogonal planes.

The general arrangement is shown in Figure 3.1. The three orthogonal planes are called the *horizontal plane*, the *frontal plane*, and the *profile plane*. For each plane the point of sight is at infinity and the projection plane is perpendicular to the projection lines.

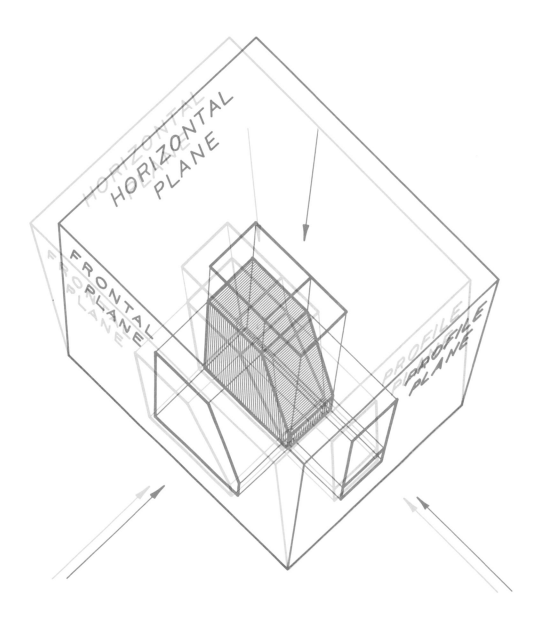

Figure 3.1 Multiview Orthographic Projection of a Solid

3.2 UNFOLDING PROJECTION PLANES

The three projection planes can be considered to be the sides of a glass or plastic box which surrounds the object. To show all three projections on one flat sheet of paper, the sides of the imaginary box can be swung into the same position as the frontal plane as shown in Figure 3.2. The resulting orthographic drawing of the object would then appear as shown in Figure 3.3. This drawing consists of three views of the object. The views are related to one another by the principle that any two adjacent views lie on planes of projection which are perpendicular to each other.

The projection on the horizontal plane is often called the "plan" or the "top view". The frontal projection may be termed the "frontal elevation" or "front view" and the profile plane may be called a "side elevation" or "side view".

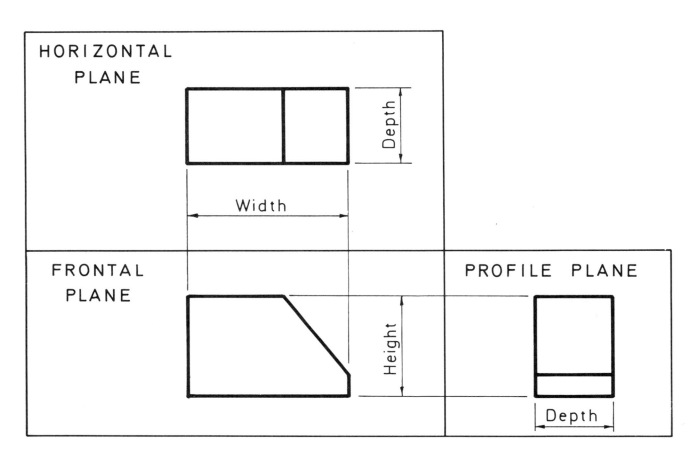

Figure 3.3 Orthographic Drawing of a Solid

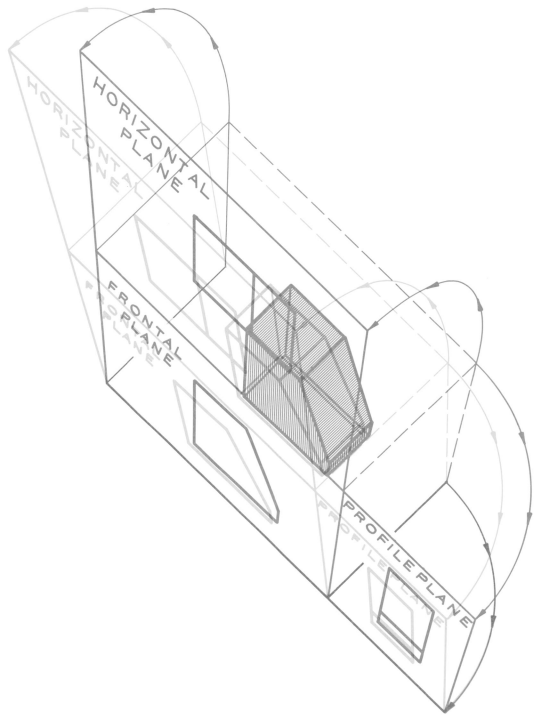

Figure 3.2 Projection Planes Folded into Frontal Plane

3.3 SPACE QUADRANTS

There are two standard systems of orthographic projection in common use. The examples in Figures 3.1 and 3.2 are drawn in *"third-angle"* projection. This is the system used generally in North America. In Europe, however, the other method, "first-angle" projection, is used. The names, first-angle and third-angle, arise from the quadrant in which the object is assumed to be placed, the quadrants being formed by the intersection of the horizontal and the frontal plane.

When the horizontal and the frontal plane are extended, they will form four quadrants in space as shown in Figure 3.4. These quadrants are designated I, II, III, and IV.

In first-angle projection, the object is assumed to lie in the first quadrant, while in third-angle projection the object is assumed to lie in the third quadrant.

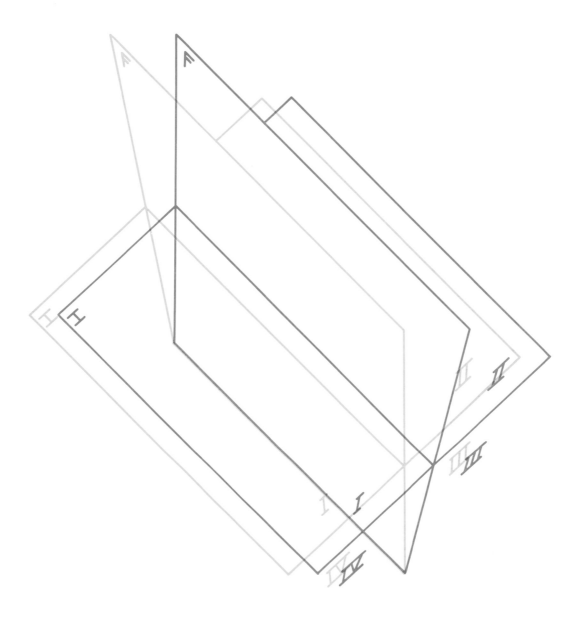

Figure 3.4 Quadrants in Space

3.4 FIRST-ANGLE PROJECTION

As an example of first-angle projection an object is shown in the first quadrant in Figure 3.5. Note that in this figure six projections are produced.

Contrary to the previous examples in which the object was behind the frontal projection plane, in first-angle projection the object is in front of the frontal plane and the front view is formed by projecting *through* the object *back* to the frontal plane. Similarly, the top view is obtained by projecting from the object *downwards* to the horizontal plane. The line of sight is considered to go *through* the object to the projection plane.

The resulting first-angle drawing of this object is shown in Figure 3.6. It is obtained by folding the projection planes into the same plane as the frontal plane, as shown in Figure 3.7.

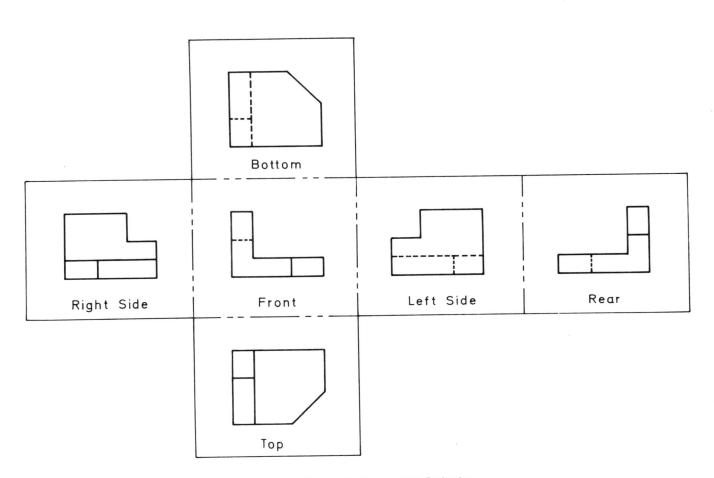

Figure 3.6 First-Angle Orthographic Projection

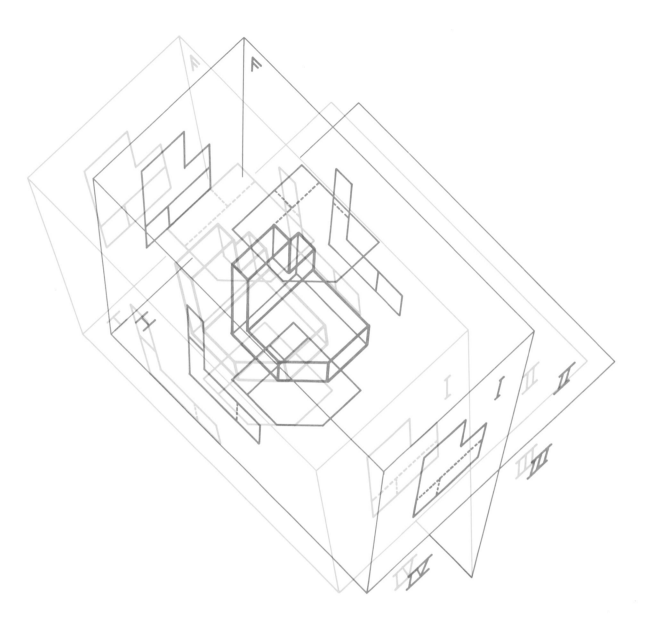

Figure 3.5 Object in First-Angle Projection

The method used to unfold the projection planes into the frontal plane in first-angle projection is shown in Figure 3.7.

Example 3.1 *First-Angle Projection*
The figures below show a pictorial sketch of an object and its first-angle projection.

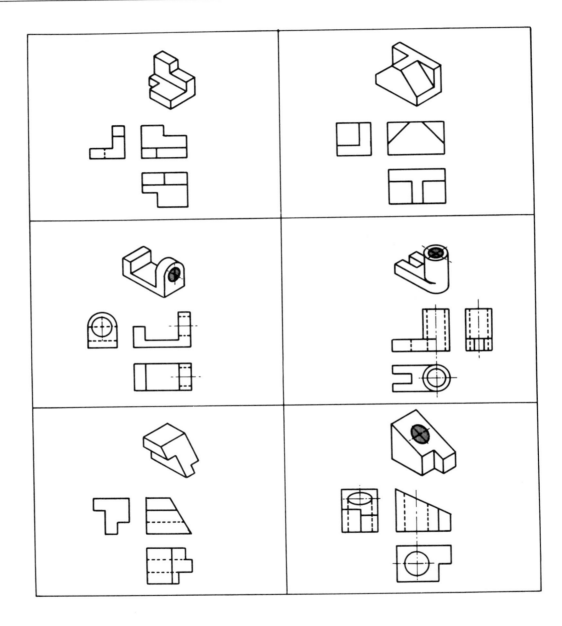

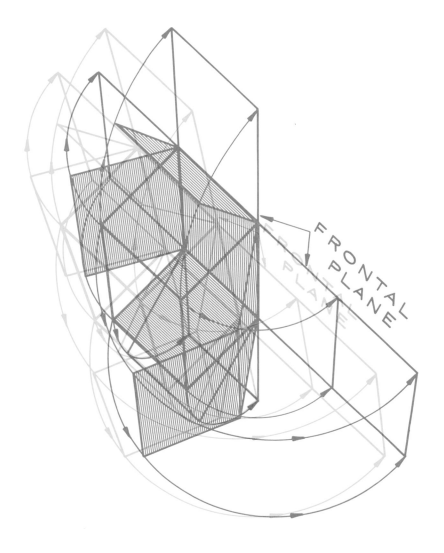

Figure 3.7 Unfolding of Planes in First-Angle Projection

3.5 THIRD-ANGLE PROJECTION

In third-angle projection, the object is considered to lie in the third quadrant and the six projections are shown in Figure 3.8. In third-angle projection, the views are produced by considering the line of sight to go through the projection plane to the object. Therefore, it can be considered that the front view is formed by projecting from the object to the frontal plane. Similarly, the top view is obtained by projecting from the object upwards to the horizontal plane.

The resulting third-angle drawing of this object is shown in Figure 3.9. It is obtained by folding the projection planes into the same plane as the frontal plane in the manner shown in Figure 3.10.

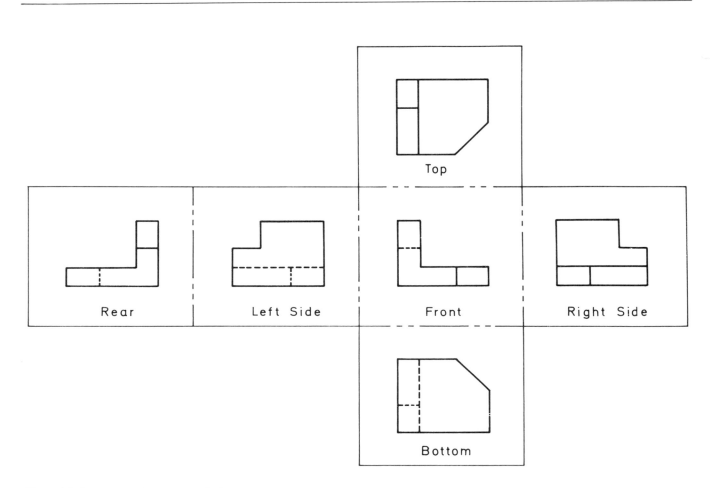

Figure 3.9 Third-Angle Orthographic Projection

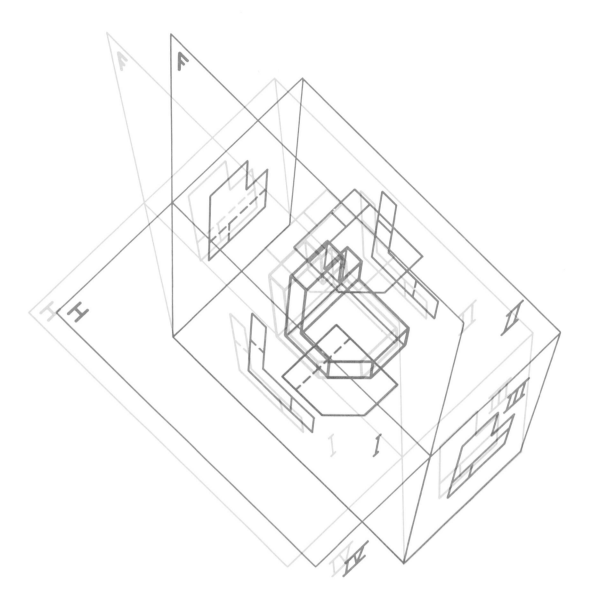

Figure 3.8 Object in Third-Angle Projection

The method used to unfold the projection planes into the frontal plane in third-angle projection is shown in Figure 3.10.

It should be noted that the resulting individual views of First-Angle Projection and Third-Angle Projection, Figures 3.6 and 3.9, are the same. The relative positions however, are changed. In third-angle projection the top view, for example, is above the front view, while in first-angle projection it is below the front view. It is apparent that to avoid confusion when reading drawings, that the same view should be in the same relative position in each drawing. This is why it is necessary to adhere to a standard projection method. In this book, all further orthographic projection drawings will be in third-angle projection.

Example 3.2 *Third-Angle Projection*
Objects from Example 3.1 shown in third-angle projection.

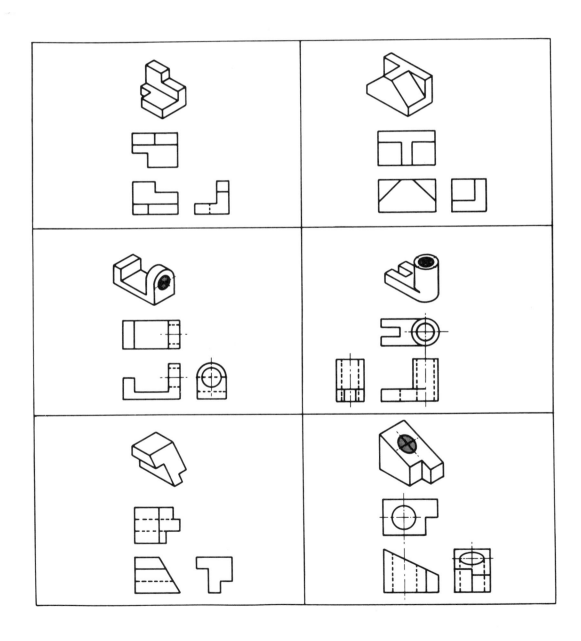

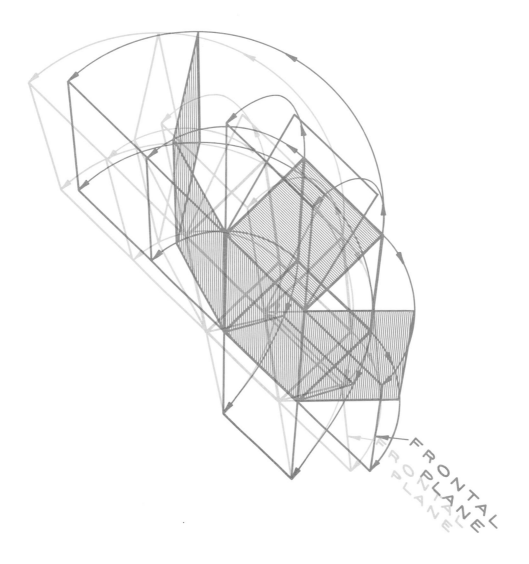

Figure 3.10 Unfolding of Planes in Third-Angle Projection

PROBLEMS

P 3.1 to P 3.4 These problems are intended to provide some experience in interpreting engineering drawings. They are useful in relating the orthographic projections of an object to a pictorial view. For each problem draw up a table as in Figure P 3.1 and fill in the numbers associated with each letter.

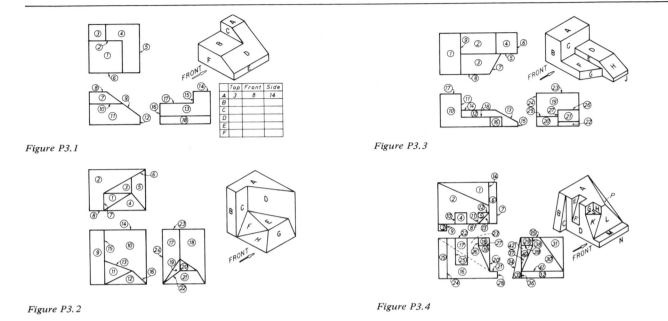

Figure P3.1

Figure P3.2

Figure P3.3

Figure P3.4

P 3.5 A house is shown pictorially in Figure P 3.5. Which is the correct front view and which is the correct side view?

Figure P3.5

P 3.6 to P. 3.8 From the pictorial views given, select the one which does not depict the same object as the others.

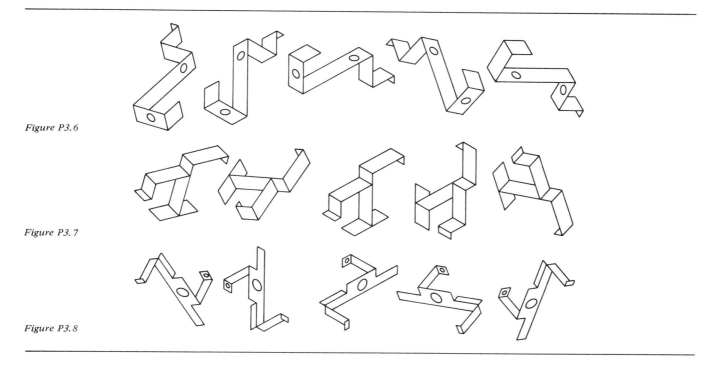

Figure P3.6

Figure P3.7

Figure P3.8

P 3.9 and P 3.10 Which block fits into block #1 in Figure P 3.9 and which into #2 in Figure P 3.10?

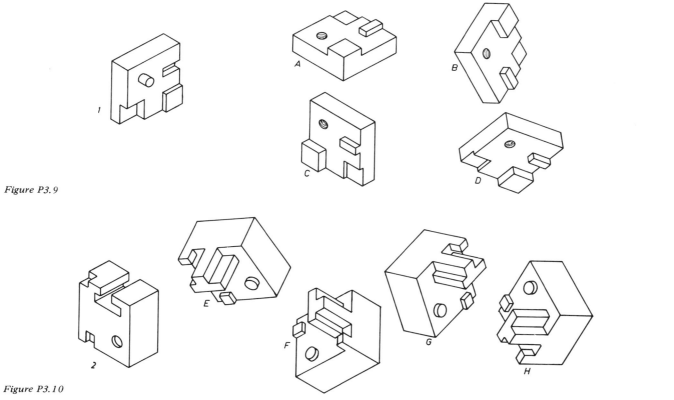

Figure P3.9

Figure P3.10

P 3.11 and P 3.12 Which are the correct top views of the objects shown pictorally?

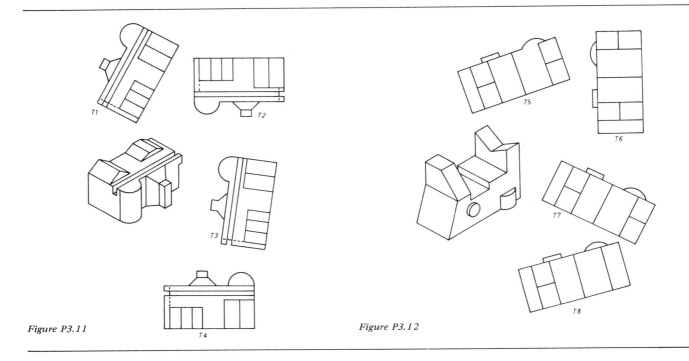

Figure P3.11

Figure P3.12

P 3.13 and P 3.14 For the object shown pictorally, which pair of orthographic third-angle projections is correct?

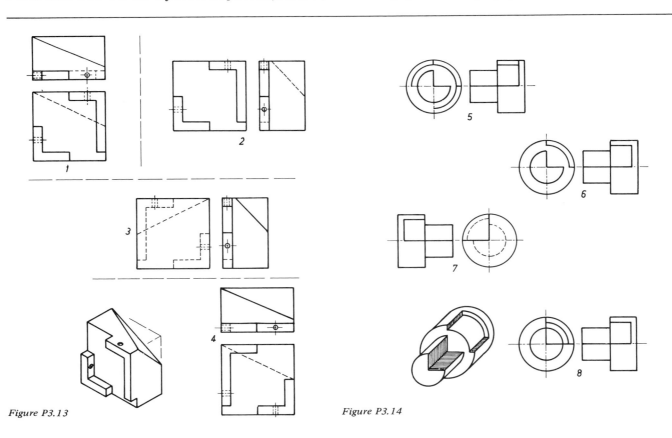

Figure P3.13

Figure P3.14

P 3.15 and P 3.16 Relate the orthographic drawings (third-angle) to the appropriate pictorial drawing for each block.

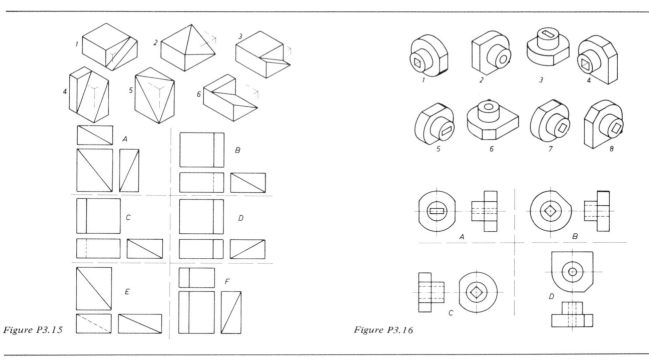

Figure P3.15

Figure P3.16

P 3.17 and P 3.18 Which is the correct pictorial drawing for the third-angle orthographic projection shown?

Figure P3.17

Figure P3.18

4. Points and Lines in Three-Dimensional Space

4.1 POINTS IN SPACE

The general procedure of representing solid three-dimensional objects on a two-dimensional sheet of paper has been described. These solid objects can be considered to be built up from a series of points and lines. Since it is so fundamental, it is necessary to consider in detail how three-dimensional points and lines may be represented on drawings by orthographic projection.

Points

In orthographic projection a point in space is represented by its projection on to three mutually perpendicular planes which may be considered to be the sides of a transparent box as shown in Figure 4.1. The actual point itself is denoted by the capital (upper-case) letter A, and its projection by lower-case letters with subscripts to indicate on to which plane the point was projected. As shown in Figure 4.1, a_H is the projection of point A on the horizontal plane, a_F is the projection on the frontal plane and a_P is the projection on the profile plane. In addition the projections of the projection lines themselves are shown. For example, $a_F T$ is the projection of line Aa_H. These projections of the projection lines are used in the actual graphical layout of points and lines.

When the sides of the box are folded out, the resulting orthographic drawing is shown in Figure 4.2. Only the reference or folding lines of the planes are shown and not the outline of the three projection planes. This is because the planes are theoretically unlimited in extent and their actual shape does not affect the drafting procedure.

It can be seen that the distance of the point *behind* the frontal plane shows up in the horizontal projection and also in the profile projection. In addition the distance *below* the horizontal plane is indicated in the frontal projection and the profile projection. This important concept should be remembered as it is frequently used in making orthographic projection drawings.

Figure 4.1 Point in Space

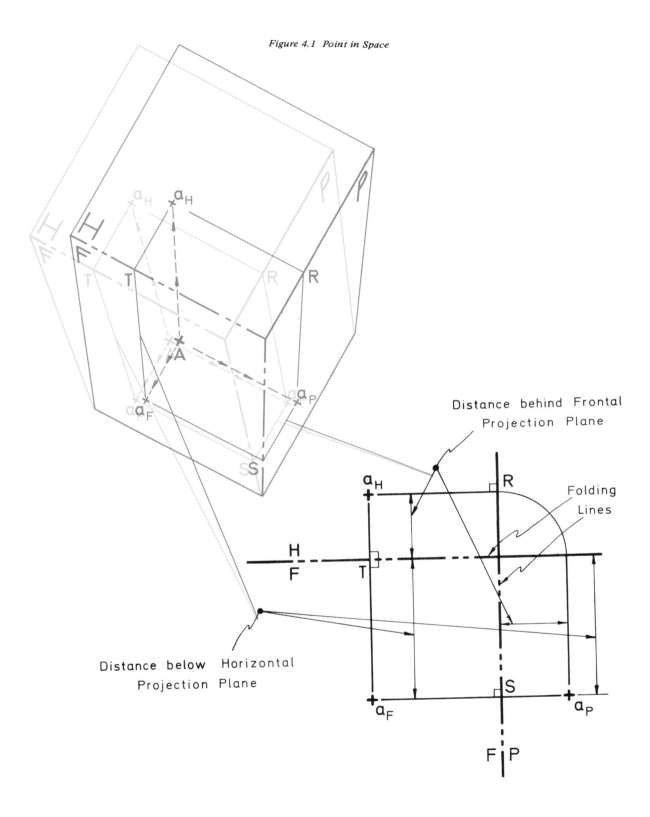

Figure 4.2 Orthographic Projection of a Point in Space

4.2 LOCATION OF A POINT IN SPACE

In order to specify the position of a point in space the familiar x, y, and z coordinates will be used. The question arises, what origin should be used for these coordinates? In practical problems one of many different origins can be specified according to the particular problem being considered. For the purposes of exercises and problems to be done by the student in connection with this text an origin for the coordinate system will be chosen at the intersection of the projection planes as shown at point 0 in Figure 4.3. Note the positive direction of the x, y, and z axes. Since we are usually working in the third quadrant, the coordinates of a point in this quadrant will have negative values. For example the point with x, y, and z coordinates of −38, −15, −20mm, is located as shown in Figure 4.3 and its orthographic projections are shown in Figure 4.4.

In Figure 4.4 the −z direction is above the origin 0 because in orthographic projection the horizontal plane is rotated into the frontal plane (see Figure 3.2). Also the −z direction appears to the right of the origin in Figure 4.4 because of the folding out of the profile plane into the frontal plane. The −x and −y directions follow directly from the arrangement shown in Figure 4.3.

Example 4.1 *Orthographic Projection—Locating a Point from Coordinates*

Given the coordinates of point B as −133, −58, −41mm. Locate the horizontal, frontal, and profile plane projections of B. Identify each point with its correct lettering. Use as an origin the point 0 given below and use a scale of 1/2 full size.

Solution

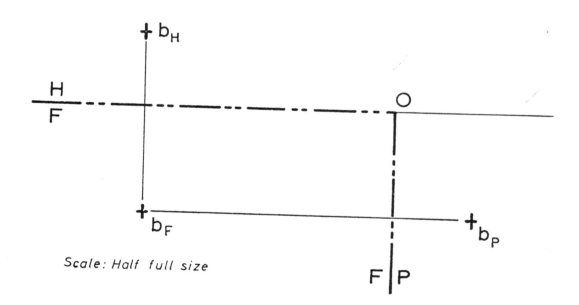

Scale: Half full size

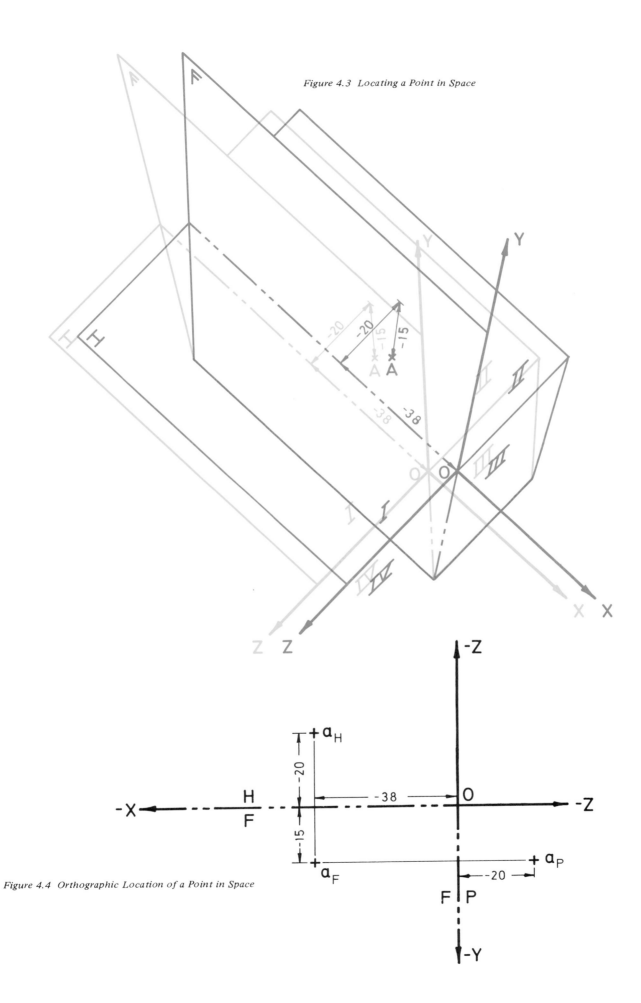

Figure 4.3 Locating a Point in Space

Figure 4.4 Orthographic Location of a Point in Space

4.3 STRAIGHT LINES IN SPACE

Since any line may be considered to be a series of points, the projection of the two end points of a straight line may be joined and will determine the position of the line in space. This is illustrated in Figure 4.5 where the line AB is shown together with its horizontal, frontal, and profile projections. In addition the projections of the dotted projectors of each end point are shown as these lines are used as construction lines in the orthographic projection views of the same line as shown in Figure 4.6.

Note that the length of the line AB in the orthographic projection of Figure 4.6 is not the correct or true length (T.L.). Each of the three views shows a somewhat shorter length. This is called foreshortening. In order to obtain a true length of a line on a projection plane, the line itself must be parallel to the plane of projection as discussed in the next section.

The folding line H/F represents one edge of the horizontal plane and also represents the horizontal projection view of the frontal plane. It is therefore an edge view of the frontal plane in the horizontal plane. Similarly the F/P line is an edge view of the profile plane in the frontal plane and also it is one edge of the profile plane.

Since a folding line represents the line of intersection of two mutually perpendicular planes it is used as a datum line to locate points by measurement as indicated in Figure 4.4. Folding lines will be indicated in this text in the following manner: ———— —— ———— —— ———— —— ————

The circular arcs in Figure 4.6 indicate that the profile projection, a_P, is the same distance behind the frontal plane as the horizontal projection a_H. Similarly the profile projection b_P is the same distance behind the frontal plane as the horizontal projection b_H.

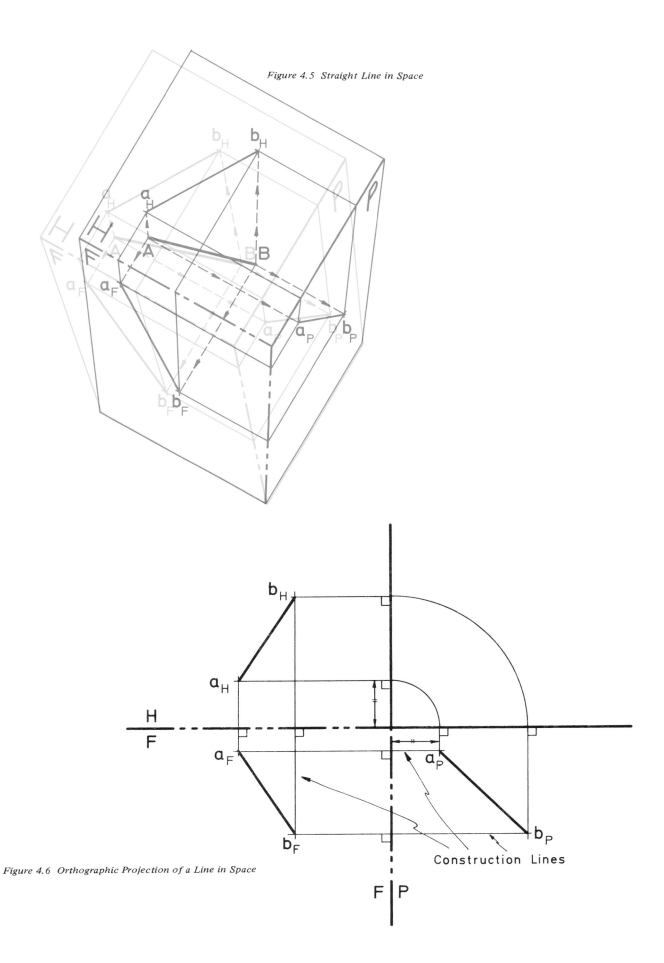

Figure 4.5 Straight Line in Space

Figure 4.6 Orthographic Projection of a Line in Space

A number of lines in space are shown in Figure 4.7. Line CD is vertical so that it projects as a point in the horizontal plane and since it is parallel to the frontal plane it will be shown in true length (T.L.) in this plane. Similarly line EF is horizontal so that its true length is shown in the horizontal projection. Line GL is parallel to both planes so that its true length is shown in both projections.

The orthographic projection of these lines on the H and F planes is shown in Figure 4.8. Note that the true length of a line is shown in the projection on that plane which is parallel to the line in space.

Note that the projection lines are shown dotted on the analglyph model of Figure 4.7. However the projections of these projection lines are not shown on this analglyph in order to simplify the model. They were shown on the previous figures, Figure 4.1 and Figure 4.5. In your orthographic drawings, these projections of the projection lines from points to the planes are used as construction lines, as indicated in Figure 4.8.

Example 4.2 *Orthographic Projections*
Draw the top, front, and right side views of the block shown in the pictorial sketch below. The location of the points A, B, and C are given as   A = $-38, -8.9, -30$ mm
B = $-10, -8.9, -13$ mm
C = $-38, -23, -13$ mm

Use the origin shown below. Label all corners of the block in each view.

Solution   Locate the projections of A, B, and C in all three views. Join AB, BC, and AC in each view. The remaining corners can be found by reference to the pictorial sketch since the block is a regular rectangular solid with one corner cut away. Note that the line $c_p a_p$ is dashed since this is a hidden line in the side view.

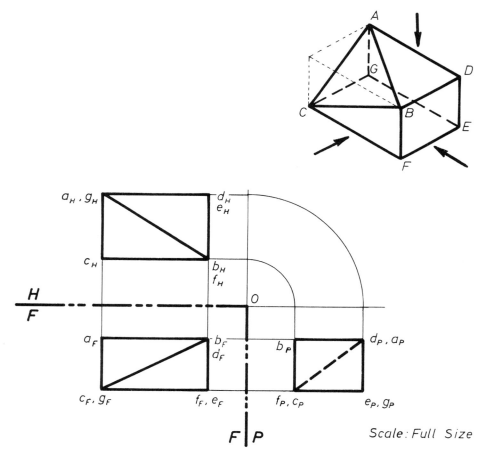

Scale: Full Size

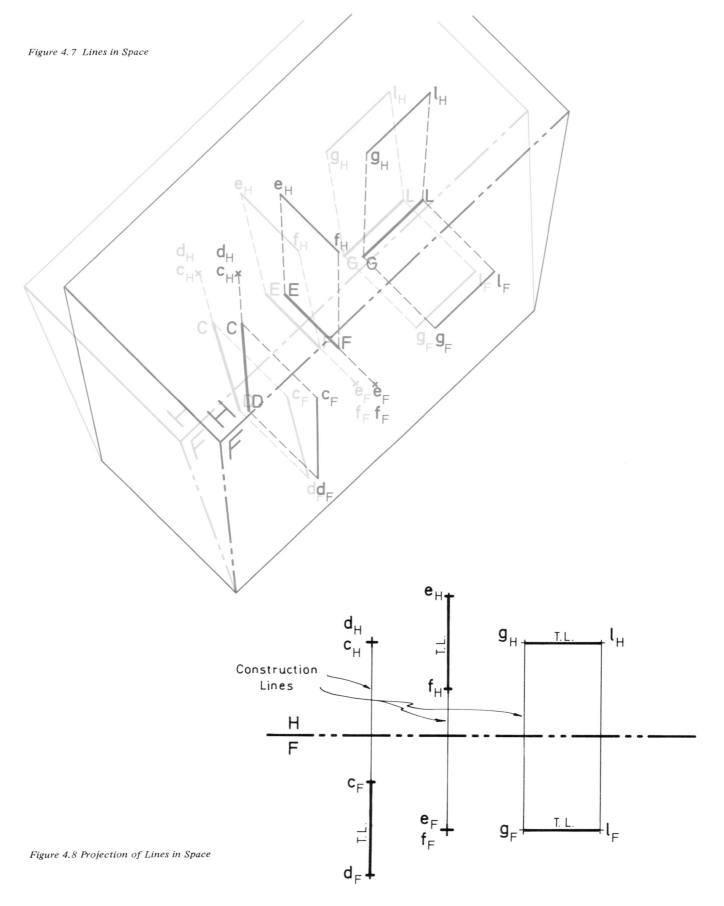

Figure 4.7 Lines in Space

Figure 4.8 Projection of Lines in Space

4.4 HORIZONTAL LINE, FRONTAL LINE, PROFILE LINE

A line parallel to the horizontal plane is called a horizontal line and is illustrated by MN in Figure 4.9. Note that the true length at this line is shown in the horizontal projection but that the frontal projection is foreshortened.

Similarly a line parallel to the frontal plane is called a frontal line as shown by RS in Figure 4.9. Again the true length of RS is shown in the frontal projection while the horizontal projection is foreshortened.

In the same manner a profile line is one that is parallel to the profile plane.

A general line not parallel to the horizontal, vertical, or profile plane is shown as TU in Figure 4.9. The true length of this line is not shown in the horizontal, frontal, or profile views but must be obtained in another manner. Two methods of obtaining a true length will be explained in the next two sections.

The three lines MN, RS, and TU are shown in Figure 4.10 in orthographic projection.

Example 4.3 *Orthographic Projection*
Draw the top, front, left, and right side projections of the object shown in the pictorial sketch below. Dimension the drawing.

Solution The folding lines and projection lines should be drawn in lightly and then erased after completing the drawing.

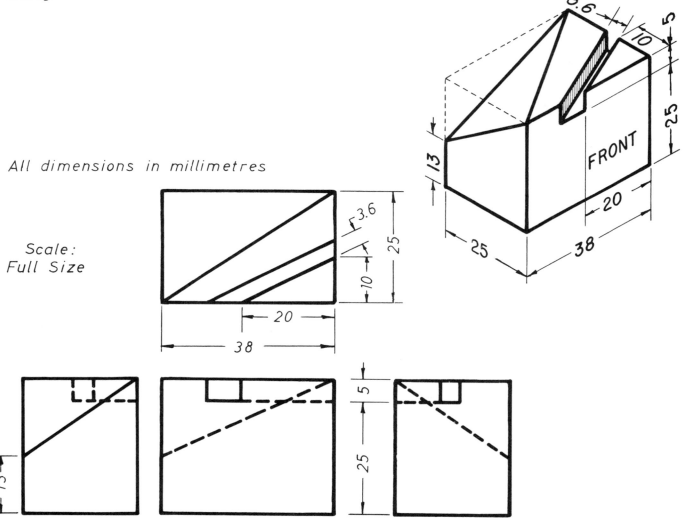

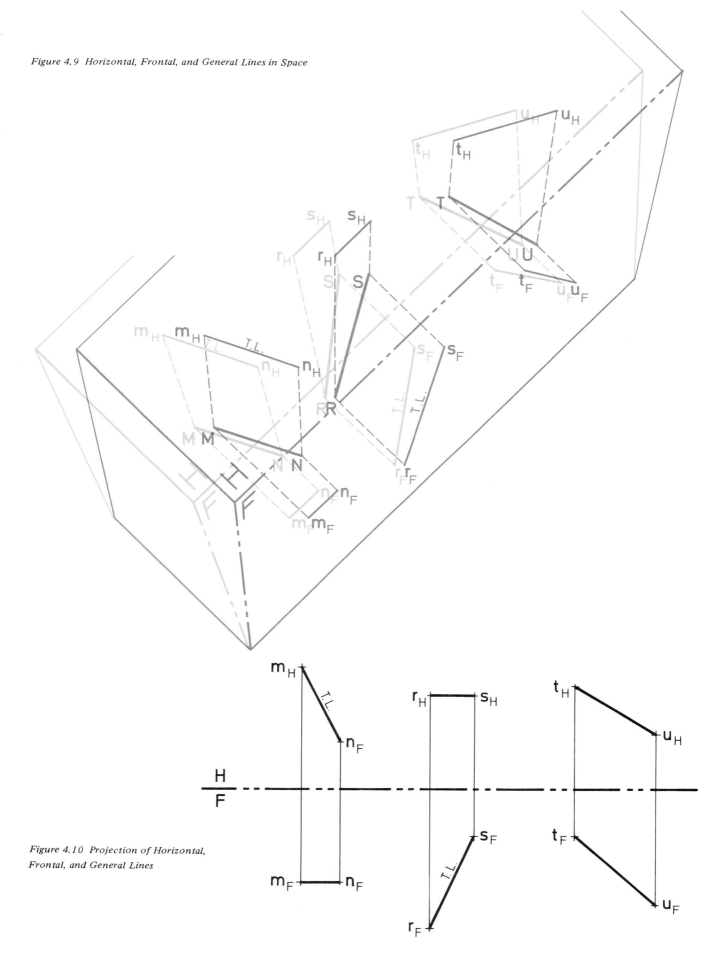

Figure 4.9 Horizontal, Frontal, and General Lines in Space

Figure 4.10 Projection of Horizontal, Frontal, and General Lines

PROBLEMS

P 4.1 The points 1 to 6 are positioned in space as shown by the horizontal and frontal projections in Figure P 4.1.
(a) Is point 1 above point 3?
(b) Is point 1 behind point 2?
(c) Is point 3 above point 4?
(d) Which is the closest point?
(e) Which two points are the same height?

P 4.2 Eight lines are located in space as shown in Figure P 4.2 by their horizontal and frontal projections.
(a) Which lines have their true lengths shown directly?
(b) Which lines are horizontal and which vertical?
(c) Are there any lines which are parallel in space?

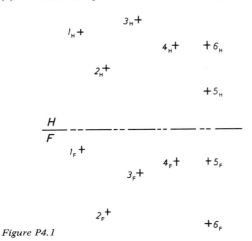

Figure P4.1

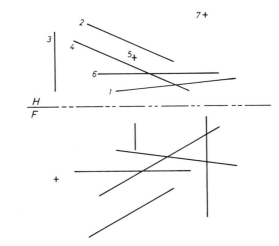

Figure P4.2

P 4.3 The orthographic projections of the objects shown have some lines missing, and some extra (incorrect) lines. Sketch the missing lines on the appropriate view, and delete the extra lines.

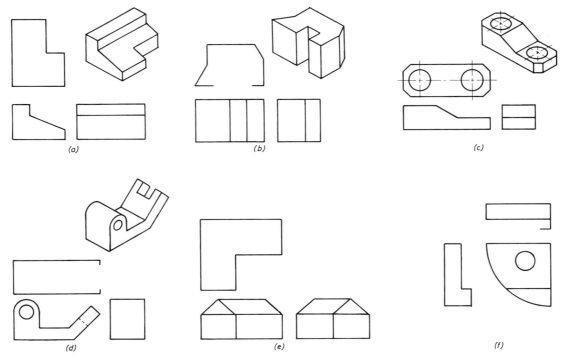

Figure P4.3

Drawing Objects from Written Descriptions

P 4.4 A symmetrical Vee-shaped block is formed from a rectangular piece 40 mm high, 75 mm wide, and 50 mm deep, by cutting a 90° Vee groove 20 mm deep into the top of the block. The groove runs the width of the block. A 5 mm diameter hole is drilled through the block in the middle of the 75 mm face and 10 mm up from the bottom. Draw top, front, and right side views. Add dimensions.

P 4.5 A symmetrical part is made of a rectangular block 80 mm square and 10 mm high, below which is a cylinder 60 mm diameter and 40 mm high. The block has four 10 mm holes each diagonally located 42 mm from the centre of the block. The cylinder has a 38 mm diameter hole drilled through its axis and through the block. Draw top and front views and dimension the drawing.

P 4.6 Each of the six faces of a 70 mm cube has a 50 mm square recess cut to a depth of 5 mm. A central hole of diameter 20 mm is drilled from top to bottom. Draw the top and front views. Add a sectional view to show the recesses. Add dimensions.

P 4.7 Consider a block 75 mm square and 60 mm high. In the top third remove and discard the left two-thirds. In the central third discard the rear two-thirds. In the bottom third discard the right one-third. Draw the top, front, and side views of the resulting object. Add dimensions.

Drawing Objects from a Pictorial View

P 4.8 to P 4.11 Draw top, front, and side views of the objects shown in the pictorial view. Dimension the views.

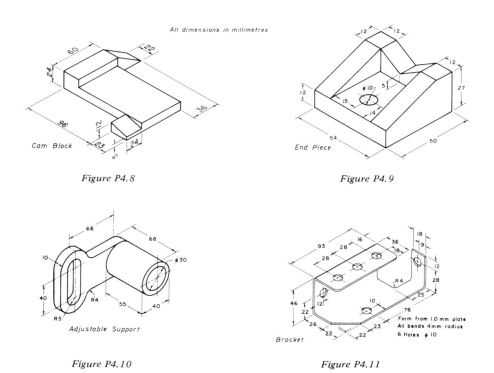

Figure P4.8

Figure P4.9

Figure P4.10

Figure P4.11

P 4.12 The orthographic projections of an object are shown in Figure P 4.12. Make an isometric sketch and answer the following questions.

(a) Which projection system, first- or third-angle, has been used?
(b) Which is the top view?
(c) Is surface A above or below surface D?
(d) Which are the two surfaces on the same level?
(e) Which is the highest surface, which the lowest?
(f) Which is the nearest surface, which the farthest?
(g) Is surface F in front of surface G?

P 4.13 For the object shown in Figure P 4.13 the surfaces B, C, D, E, F, and G are all horizontal.

(a) Is the projection first- or third-angle?
(b) Is surface B higher than surface F?
(c) Is surface D higher than surface F?
(d) Which three surfaces are at the same height?
(e) Which surface is the highest?

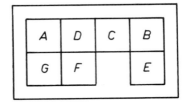
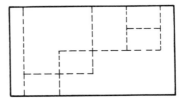

Figure P4.13

P 4.14 For the object shown indicate the lettered features on the top view.

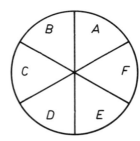

Figure P4.14

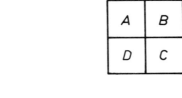
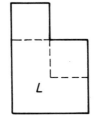
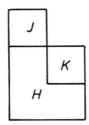
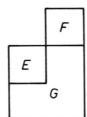

Figure P4.12

P 4.15 Four pairs of related projection views are given for the object shown pictorially in the upper left-hand position. For each pair of views sketch the view indicated by the arrows.

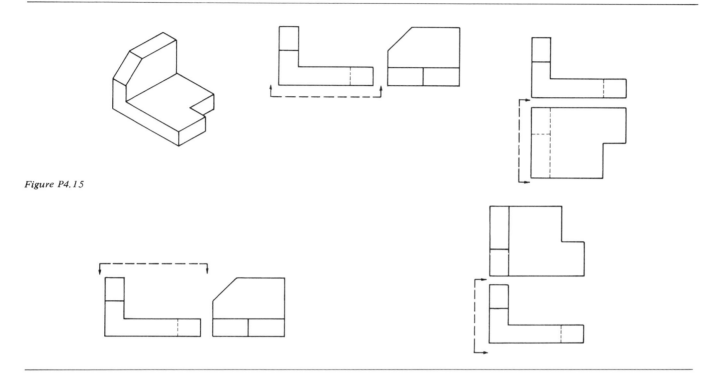

Figure P4.15

P 4.16 Three views and a section are shown for a machine part.

(a) Is surface A in front of surface B?
(b) What is the distance between surface A and surface B?
(c) Determine the distance indicated by the letter C.
(d) Determine the distance indicated by the letter D.
(e) What is the height of the surface E above the base?
(f) What is the distance indicated by the letter F?
(g) What is the distance between the sectioned surface G and the surface H?

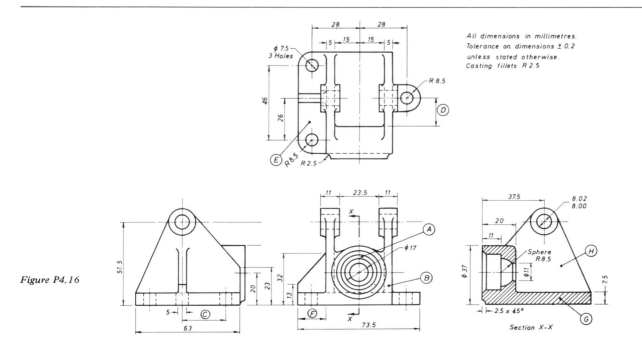

Figure P4.16

5. Auxiliary Planes

5.1 AUXILIARY PLANE AND TRUE LENGTH

So far we have considered three mutually perpendicular planes, the horizontal, the frontal, and the profile planes and the views projected on to these planes. The line AB and its projections on these three planes is shown in Figure 5.1. Since AB is not parallel to any of these planes, none of the projections shows the true length of the line. The orthographic projections are shown in Figure 5.2.

An auxiliary plane may be defined as a plane which is at any orientation we may choose and the projection on to that plane is called an auxiliary view.

In order to determine the true length of a line we select an auxiliary plane, oriented so that it is parallel to the line in space, and obtain an auxiliary view by projection on to that plane. Such a plane is shown in Figure 5.3 drawn parallel to the line AB and its horizontal projection. We will denote this plane by the number 1 as shown. The orthographic drawing is shown in Figure 5.4, the true length being found on the auxiliary view. It is important to remember that the distance of a_1 behind the H/1 line is the same distance that a_F is behind the H/F line. Similarly for b_1 and b_F. See Example 5.2 on page 53.

5.2 BEARING OF A LINE

The direction of a line relative to a compass heading, usually North, is called the bearing of the line. The bearing is determined from the horizontal projection since a compass is held in a horizontal plane. As indicated in Figure 5.2 the bearing of AB is N 37° E.

5.3 SLOPE OF A LINE

The slope of a line is the inclination that the line has with respect to the horizontal plane. It can only be seen in a projection in which the line is a true-length view in a plane which is perpendicular to the horizontal plane. This is illustrated in Figure 5.4 which shows the angle $\theta_H = 45.5°$ which is the angle between the true length of the line and the horizontal plane.

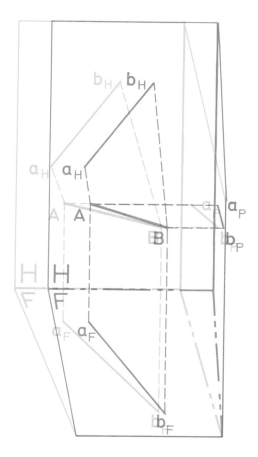

Figure 5.1 Line in Space

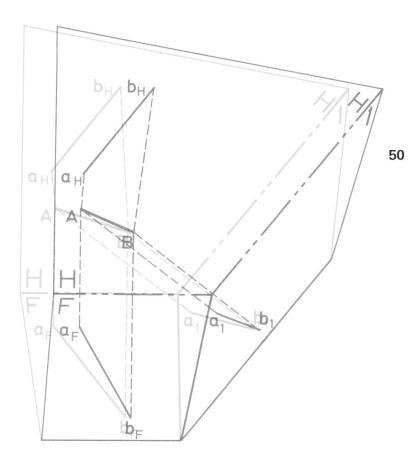

Figure 5.3 Line in Space—Auxiliary Plane

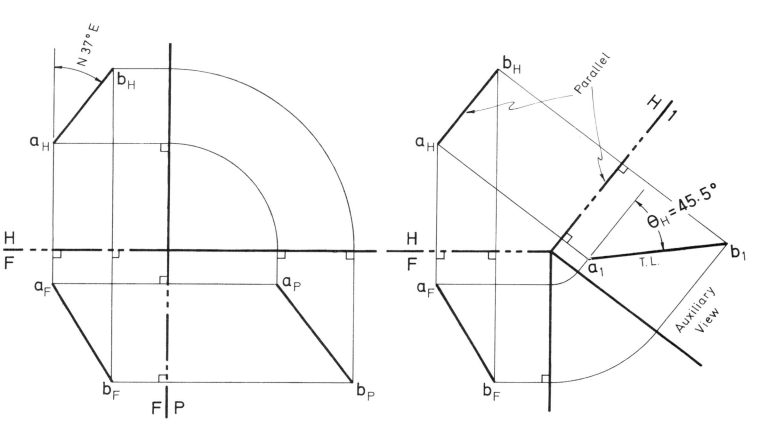

Figure 5.2 Projection—Line in Space

Figure 5.4 Projection on Auxiliary Plane

5.4 TRUE LENGTH OF A LINE BY THE ROTATION METHOD

In the system of orthographic projection so far discussed we have assumed that the object is fixed in position, and we have obtained the projections from different points of view to develop the projections on to the horizontal, frontal, profile, and auxiliary planes. In the rotation method we permit the line to rotate about an axis so that it becomes parallel to one of the existing projection planes. In this manner the true length and inclination will then appear as a projection on that plane, and an auxiliary plane is not required. See also Chapter 11, Rotation.

When a line is rotated about an axis through one end it will generate a cone. This is shown in Figure 5.5 and in Figure 5.6. The true length of AB can be seen in the two positions in the frontal plane when it becomes parallel to this plane, positions AB_1 and AB_2. The inclination of the line with the horizontal plane, θ_H, is also shown. Note that in the projection view, Figure 5.6, the end b_H is rotated about a_H and the frontal projection b_F generates a straight line since it is the view of the flat bottom of the cone.

Of course it is not necessary to draw the complete cone in using this method. Also the line can be rotated about a horizontal axis as shown in Figures 5.7 and 5.8 in order to find the true length and its inclination with the frontal plane as indicated.

Example 5.1 *True Length and Slope of a Given Line by Rotation*

The line AB is defined from the given origin 0 as:
A = −28, −28, −21 mm;
B = −20, −8.5, −8.5 mm.

Find the true length and the slope that the line makes with the horizontal.

Solution Rotation may be made either about end A or end B.

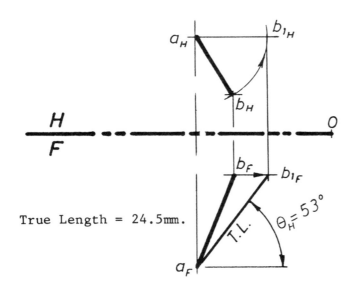

True Length = 24.5mm.

Rotation about A

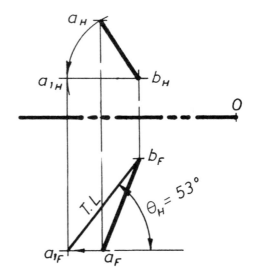

Rotation about B

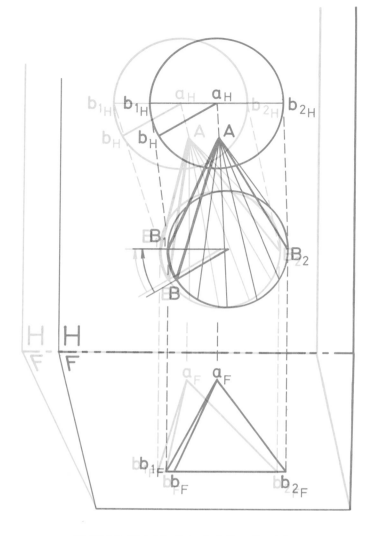

Figure 5.5 True Length by Rotation—Frontal

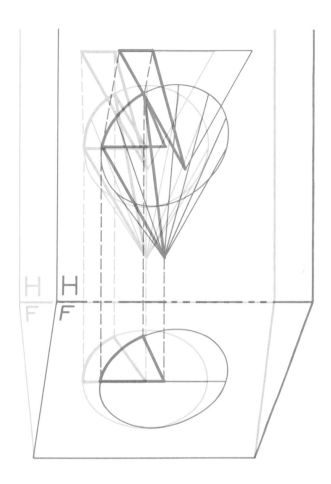

Figure 5.7 True Length by Rotation—Horizontal

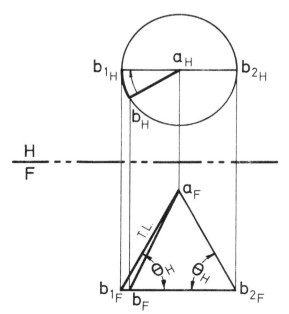

Figure 5.6 Projection of Rotation Method

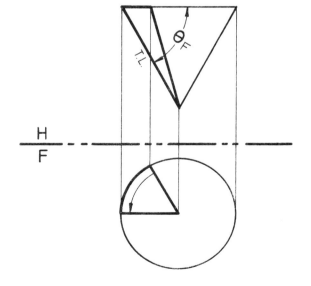

Figure 5.8 Projection of Rotation Method

5.5 AUXILIARY INCLINED PLANES AND THE POINT VIEW OF A LINE

An auxiliary inclined plane, sometimes called a double auxiliary plane, is one in which the observer's line of sight is neither vertical nor horizontal. It is called double auxiliary because it is adjacent to an auxiliary plane. At this stage we will consider the use of an auxiliary inclined plane to provide the point view or end view of a line.

In order to obtain a view in which a line becomes a point we must look along the line. That is, the projection plane must be perpendicular to the line. This means that the projection plane must be perpendicular to the *true length* of a line in order for it to project as a point. In Figure 5.9 the line AB is shown in true length in the auxiliary plane 1.

An inclined auxiliary plane 2 is then added perpendicular to the true length to form a point view of AB. The orthographic projection is shown in Figure 5.10.

A fundamental principle of orthographic projection, that adjacent projection planes are at right angles to each other, is illustrated in Figure 5.9. For example the planes H and F are perpendicular, planes H and 1 are perpendicular, and planes 1 and 2 are perpendicular to each other. Note that the F plane and plane 1 are not perpendicular to each other, and also the H plane and plane 2 are not perpendicular to each other. These planes also are not adjacent as shown in the orthographic projection of Figure 5.10.

Example 5.2 *True Length of a Line by Use of an Auxiliary Plane*
Find the length of AB, A = −28, −23, −16;
B = −20, −3.5, −3.5 mm

Solution The true length may be obtained using an auxiliary plane either adjacent to the frontal plane or adjacent to the horizontal plane.

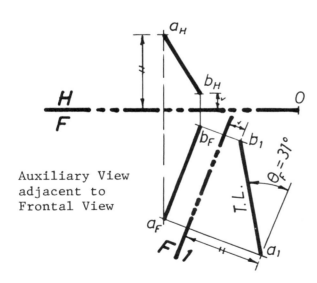

Auxiliary View adjacent to Frontal View

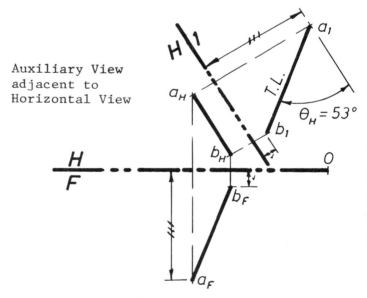

Auxiliary View adjacent to Horizontal View

True length = 24.5mm

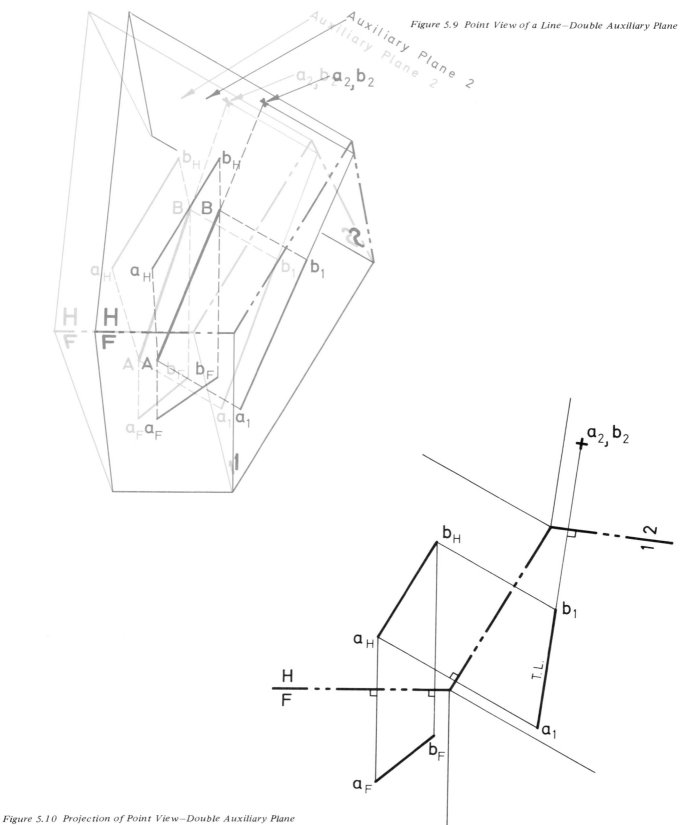

Figure 5.9 Point View of a Line—Double Auxiliary Plane

Figure 5.10 Projection of Point View—Double Auxiliary Plane

54

PROBLEMS

All dimensions in millimetres unless otherwise stated.

P 5.1 and P 5.2 Find the true length of the line shown. Project on to an auxiliary plane off the horizontal plane. Check by using an auxiliary plane off the frontal plane. Use the rotation method as a further check. Report the angle the line makes with the horizontal plane and with the frontal plane.

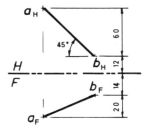

Figure P5.1

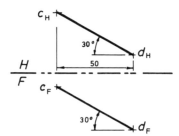

Figure P5.2

P 5.3 Draw top, front, and side views of the angle bracket shown in Figure P 5.3. Also draw an auxiliary view showing the bent down portion only in true size.

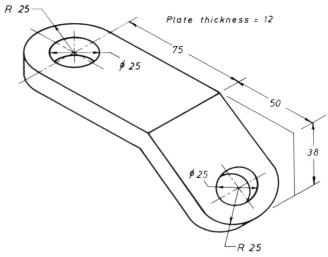

Figure P5.3

P 5.4 An item of electronic research equipment consists of a wire probe AB set in an insulated box as shown in Figure P 5.4. What is the length of AB and its angle of inclination to the top of the box? Also find the angle of inclination with each of the two sides.

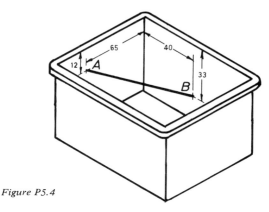

Figure P5.4

P 5.5 Given the line CD and the partial location of points A and B as shown in Figure P 5.5, draw a frontal line rising up from B, 60 mm long and ending at point E on CD. Also draw a horizontal line AF running forward from F, 40 mm long and terminating on CD at point F. Also draw the profile view of all three lines.

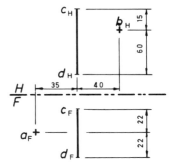

Figure P5.5

P 5.6 For the part shown in Figure P 5.6 draw top, front, and left side views. Also draw an auxiliary view to show the true shape of face A. Dimension the drawing.

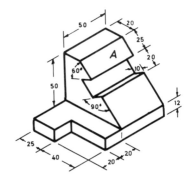

Figure P5.6

P 5.7 The horizontal and frontal projections of a triangle EFG are shown in Figure P 5.7. Draw the profile view of the triangle. In all three views locate the following points: A on side EF 20 mm to the right of E; B on FG 40 mm above G; C on EG 10 mm behind G; D lying in the extended plane of EFG, d_H is given.

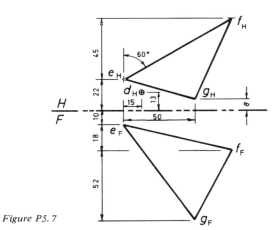

Figure P5.7

P 5.8 A communications tower is braced by three steel guy wires as shown in Figure P 5.8. Find the length of each wire. Choose a suitable scale for your drawing.

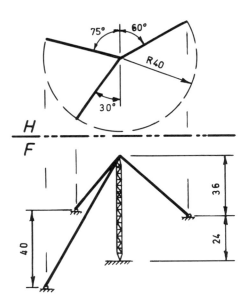

Figure P5.8

P 5.9 A cable supported stadium roof is outlined in Figure P 5.9. Find the length of cables AB, AC, and AD. What is the slope angle of cable AB? Choose a suitable scale for your drawing.

Dimensions in metres

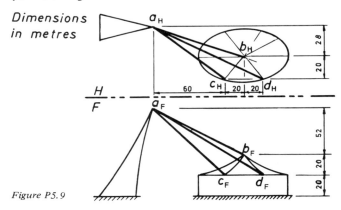

Figure P5.9

P 5.10 A steel frame made to hoist a load from point A is constructed as shown in Figure P 5.10. Point B is in the middle of AC. Member BD has a bearing of N 17° E. What is the true length of BD? What is the slope of AG? When member AC appears as a point what is the angle between each set of side braces?

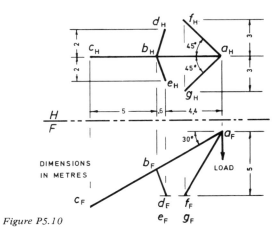

Figure P5.10

P 5.11 Determine whether the lines PQ and RS intersect or not.

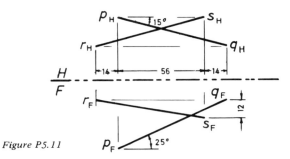

Figure P5.11

6. Perpendicular Lines

6.1 INTERSECTING PERPENDICULAR LINES IN SPACE

The 90° right angle between two intersecting perpendicular lines will be seen in any view in which one of the lines appears as a true length. Another condition which proves perpendicularity is the view in which one of the lines appears as a true length and the other as a point view on the true length.

Two sets of intersecting perpendicular lines are shown in Figure 6.1. Lines AB and CD both appear as true lengths in the horizontal projection and the 90° angle between them can be seen. In the frontal projection, line AB is shown as a point and the line CD is a true length. Either of these views prove perpendicularity.

Lines EF and GH are in such a position in space that GH is foreshortened in the horizontal projection, $g_H h_H$, but since EF appears as a true length, $e_H f_H$, the 90° angle can be seen in the horizontal projection. In the frontal projection the line EF is seen as a point view which again proves perpendicularity since $g_F h_F$ is a true length.

Lines JK and LM are not perpendicular even though the projection of JK on the frontal plane, $j_F k_F$, appears as a point on $\ell_F m_F$. However, $\ell_F m_F$ is not a true length and thus this view does not prove perpendicularity. The orthographic projections are shown in Figure 6.2.

Figure 6.1 Intersecting Perpendicular Lines

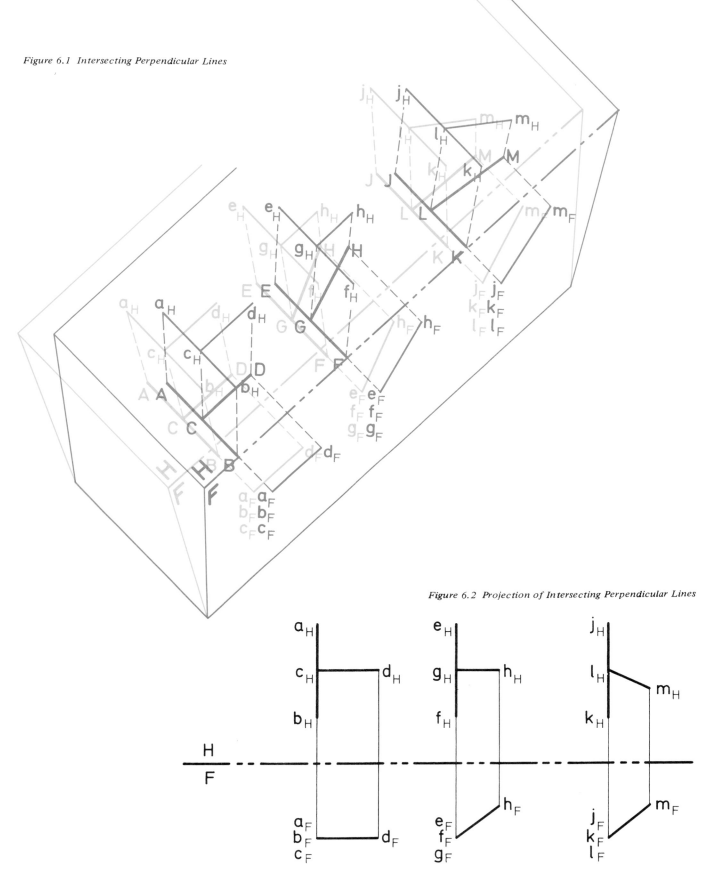

Figure 6.2 Projection of Intersecting Perpendicular Lines

6.2 TWO INTERSECTING PERPENDICULAR LINES—GENERAL CASE

A more general situation of two intersecting perpendicular lines is shown in Figure 6.3. Note that in this case the 90° angle between the two lines cannot be seen in the standard horizontal and frontal projections. This is because neither of the lines appears as a true length in these projections. However the auxiliary plane 1, chosen to be parallel to line CD will show $c_1 d_1$ as a true length and the 90° angle will be seen in this projection as illustrated in the orthographic projections of Figure 6.4.

Example 6.1 *Perpendicular Lines*

Given the two lines AB and CD defined below, determine if these lines are perpendicular to each other.

A = −38, −9, −9 mm
B = −15, −19, −34 mm
C = −12, −8, −18 mm
D = −29, −18, −38 mm

Solution Perpendicularity can be seen in any view in which one of the lines appears in true length. Two solutions are shown below. The solution on the left was developed by taking a plane parallel to $a_F b_F$ so that $a_1 b_1$ is a true length. The solution on the right was made by taking a plane parallel to $c_F d_F$ so that $c_1 d_1$ is a true length. Solutions could have been obtained by taking auxiliary planes adjacent to the horizontal plane.

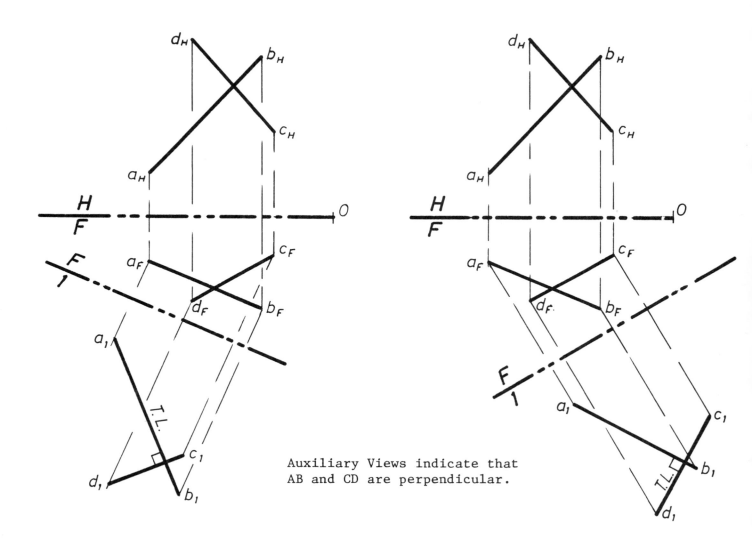

Auxiliary Views indicate that AB and CD are perpendicular.

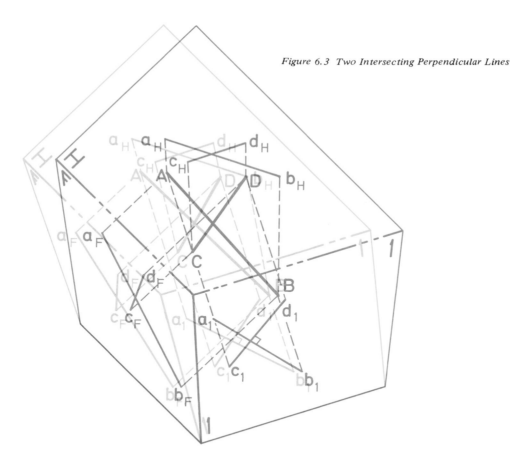

Figure 6.3 Two Intersecting Perpendicular Lines

Figure 6.4 Projection of Two Intersecting Perpendicular Lines

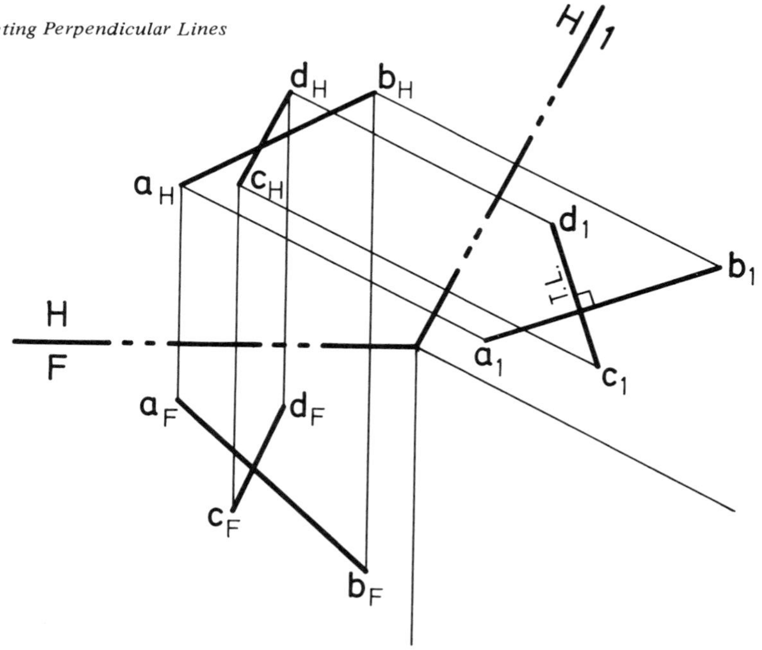

6.3 NON-INTERSECTING PERPENDICULAR LINES

Non-intersecting, non-parallel lines are called *skew lines*. Two such lines that also have a direction perpendicular to each other are shown in Figure 6.5. The 90° between them cannot be seen in either the horizontal or the frontal projection but is visible in the auxiliary plane 1 which is parallel to the line CD so that $c_1 d_1$ is a true length. (Figure 6.6) The perpendicular distance between any two skew lines will be considered later.

Example 6.2 *Construction of a Perpendicular to a Given Line*

Given the line CD construct a line through the point A perpendicular to CD.

A = −45, −42, −41 mm
C = −43, −4, −45 mm
D = −26, −33, −67 mm

Solution The perpendicular must be constructed on a true length view of CD. Therefore draw the auxiliary view in plane 1 parallel to $c_H d_H$. Drop a perpendicular from a_1 to $c_1 d_1$. This line can terminate at any convenient point x_1. To obtain x_H project from x_1 back to the horizontal plane by drawing a line from x_1 perpendicular to the folding line H/1. Locate x_H at any convenient point on this line. Project from x_H down to the frontal plane to find x_F as shown.

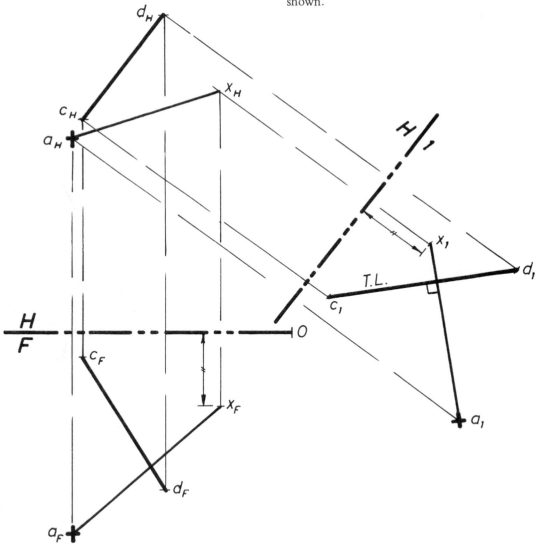

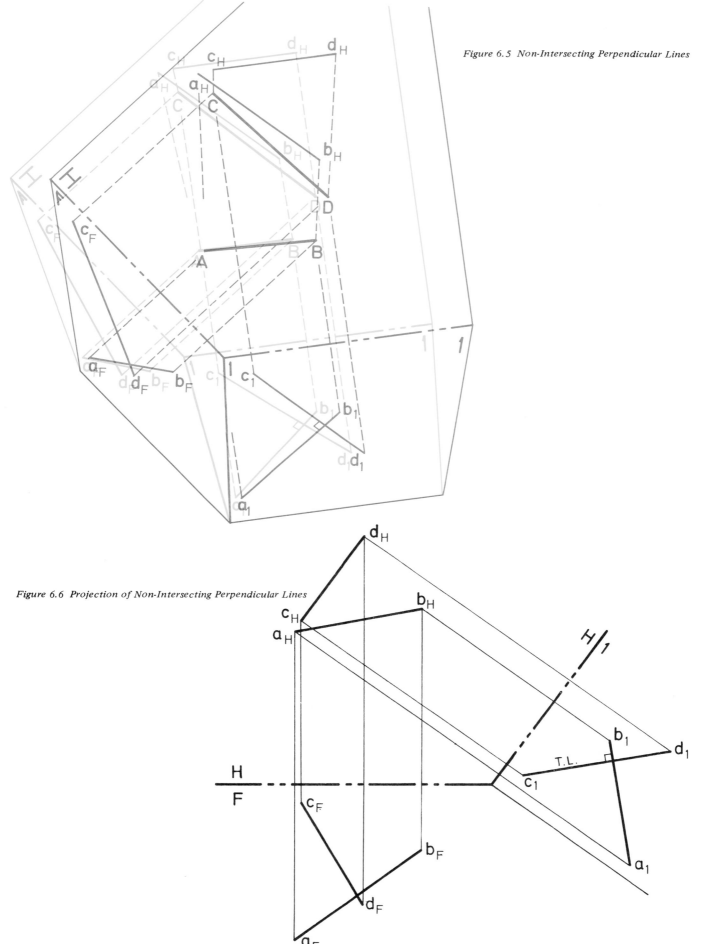

Figure 6.5 Non-Intersecting Perpendicular Lines

Figure 6.6 Projection of Non-Intersecting Perpendicular Lines

PROBLEMS

Dimensions in millimetres unless otherwise specified.

P 6.1 Determine if the two lines AB and CD are perpendicular to each other.

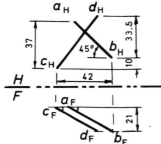

Figure P6.1

P 6.2 Determine if the two lines EF and GH are perpendicular to each other.

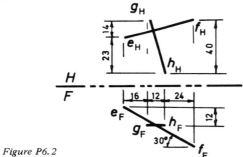

Figure P6.2

P 6.3 Draw H and F views of line AB which is perpendicular to CD. B is to lie on CD.

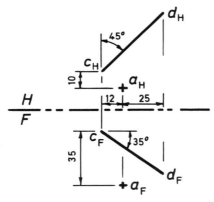

Figure P6.3

P 6.4 Draw H and F views of line PQ which is perpendicular to RS, Q being on line RS.

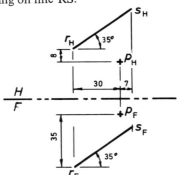

Figure P6.4

P 6.5 Draw and H and F views of lines AB and DC which are perpendicular to line BC. Positions a_H only and d_F only are given.

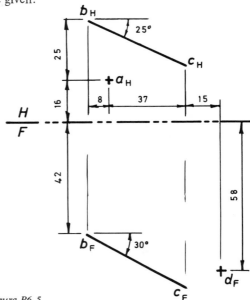

Figure P6.5

P 6.6 From point E construct line EF intersecting line GH at F and making an angle of 90° with GH.

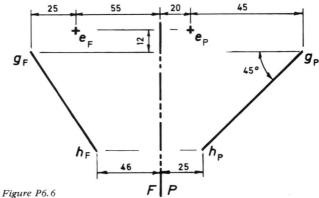

Figure P6.6

P 6.7 Given the vertex V and the centre of the base at A of a right pyramid as shown in Figure P 6.7. The base is a square with diagonals 80 mm long. One corner C of the base is behind A and 25 mm below A. Complete the horizontal, frontal, and auxiliary views necessary to the construction.

P 6.8 Given the frontal and left profile projections of the axis AB of a right circular cone, the vertex being at A. The base of the cone is 50 mm in diameter. Draw the frontal view of the cone and auxiliary views required in the construction. It will be convenient to project from the profile view.

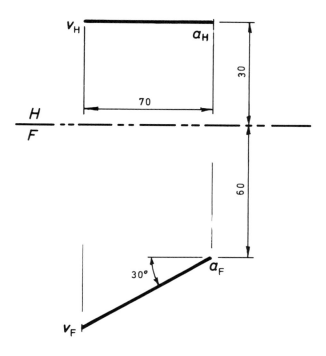

Figure P6.7

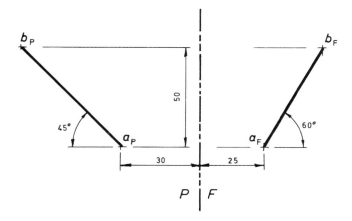

Figure P6.8

7. Parallel Lines

7.1 PARALLEL PROJECTIONS

Parallel lines in space will appear as parallel projections in all views except when viewed from two positions:
a) when the lines appear as end views or points, and
b) where the lines appear one behind the other.

It is apparent that when two lines appear as points this is proof that the lines are parallel. On the other hand, if two lines appear one behind the other, it cannot be stated that they are parallel unless another view shows them to appear as points or to be parallel.

Two parallel lines are shown in Figure 7.1 in various projection planes. Starting with the standard horizontal and vertical projection views, we create an auxiliary plane 1 to find their true length and then another auxiliary inclined plane 2 at right angles to the true lengths to show the end or point views. The corresponding orthographic projections are shown in Figure 7.2.

Example 7.1 *Line Parallel to a Profile Line*

Given the profile line AB, draw a line CD through C parallel to AB.

A = −20, −32, −10 mm
B = −20, −11, −23 mm
C = −28, −28, −8 mm

Solution The left figure shows the line AB and the point C. The required line will have parallel projections. That is $c_H d_H$ will be parallel to $a_H b_H$ and $c_F d_F$ will be parallel to $a_F b_F$. However we do not know where the terminus D should be. A profile view must be drawn as shown in the right-hand figure. Then the slope of AB is visible and $c_P d_P$ can be drawn parallel to $a_P b_P$ to any convenient length.

The true distance between the two lines can be obtained from an auxiliary view showing the lines as points, as indicated in plane 1.

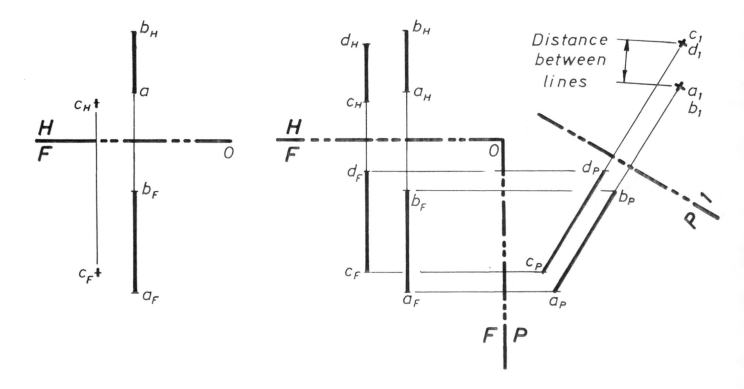

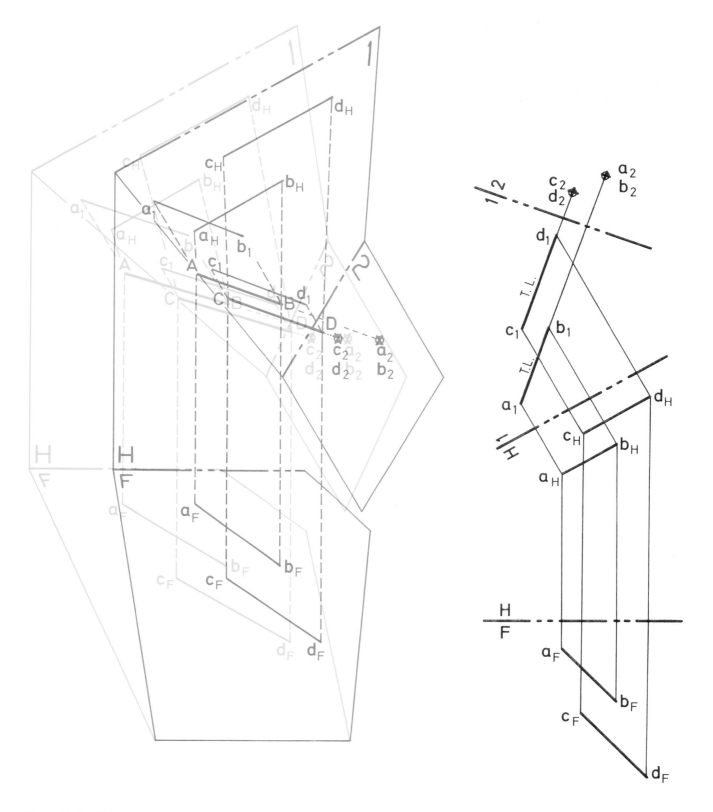

Figure 7.1 Parallel Lines

Figure 7.2 Projection of Parallel Lines

8. Plane Surfaces

8.1 GENERAL FORMATION

A plane surface is one which is not curved or warped. We have considered various *projection* planes, horizontal, profile, auxiliary, and inclined or double auxiliary. We are now concerned with the representation of general plane surfaces in space by projection on to these different projection planes.

A plane may be defined by several different conditions as illustrated in Figure 8.1.

1) Two intersecting lines AB and CD form a plane as shown in Figure 8.1(a).
2) Two parallel lines EF and GH form a plane. Figure 8.1(b).
3) A straight line, JK, and a point, P, determine a plane, Figure 8.1(c). This situation can be transformed to the two parallel lines, case 2, by drawing a parallel line through the point P.
4) Three non-collinear points, R, S, and T, Figure 8.1(d), form a plane. This case can be transformed to the case (1) of two intersecting lines, RS and ST as shown.

The orthographic projections of these planes are shown in Figure 8.2.

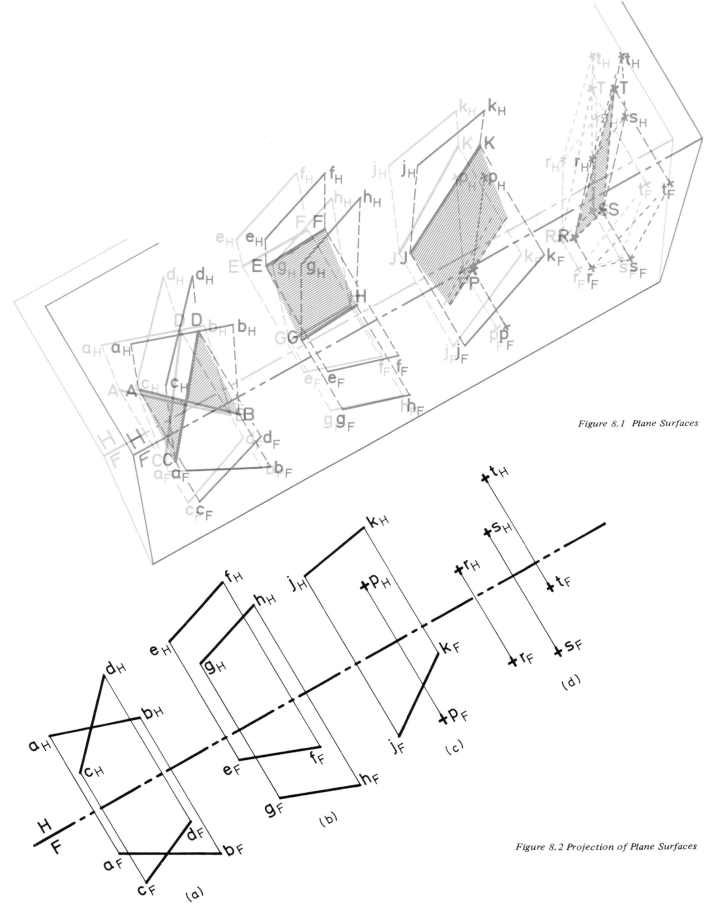

Figure 8.1 Plane Surfaces

Figure 8.2 Projection of Plane Surfaces

8.2 EDGE VIEW AND TRUE SLOPE ANGLE OF AN OBLIQUE PLANE

An oblique plane may be defined as one which is inclined to all three principal planes of projection.

An oblique plane may be seen in an edge view by looking along the true length of any line in the plane. In this situation the line appears as a point. In other words a point view of a line lying in a plane will show the plane as an edge view.

As an illustration, it is required to find the edge view of the given plane ABC shown in Figure 8.3. We have to obtain the true length of a line in this plane so for convenience choose the horizontal line BX as shown in the orthographic view of Figure 8.4. This line is chosen because the true length will appear in the horizontal projection.

An auxiliary plane, 1, perpendicular to the true length of BX will provide the required edge view (E.V.) as shown.

The true slope angle, called the dip angle, is the angle between a plane and the horizontal plane. This angle will be seen in a projection plane which is not only adjacent to the horizontal plane but also is one in which the oblique plane appears in an edge view.

Example 8.1 *Intersection of Two Planes*
Given the planes ABC and DEFG determine their line of intersection.

Solution If two planes intersect, their line of intersection may be found by obtaining an edge view of one of the planes.

Obtain an edge view of plane ABC by taking auxiliary view 1 along the true length $b_H x_H$. In view 1 it can be seen that plane ABC intersects plane DEFG at v_1 and w_1. v_1 is on line GF and w_1 is on line DE. Project v_1 and w_1 back from view 1 to the horizontal and frontal projections.

The required line of intersection (extended) is the line VW.

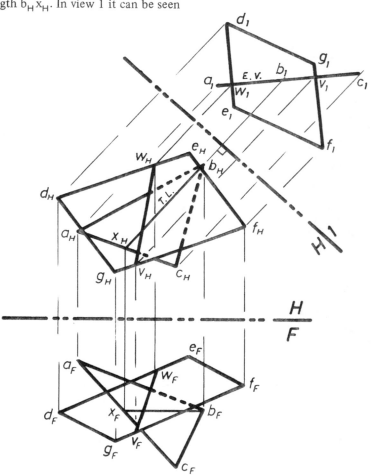

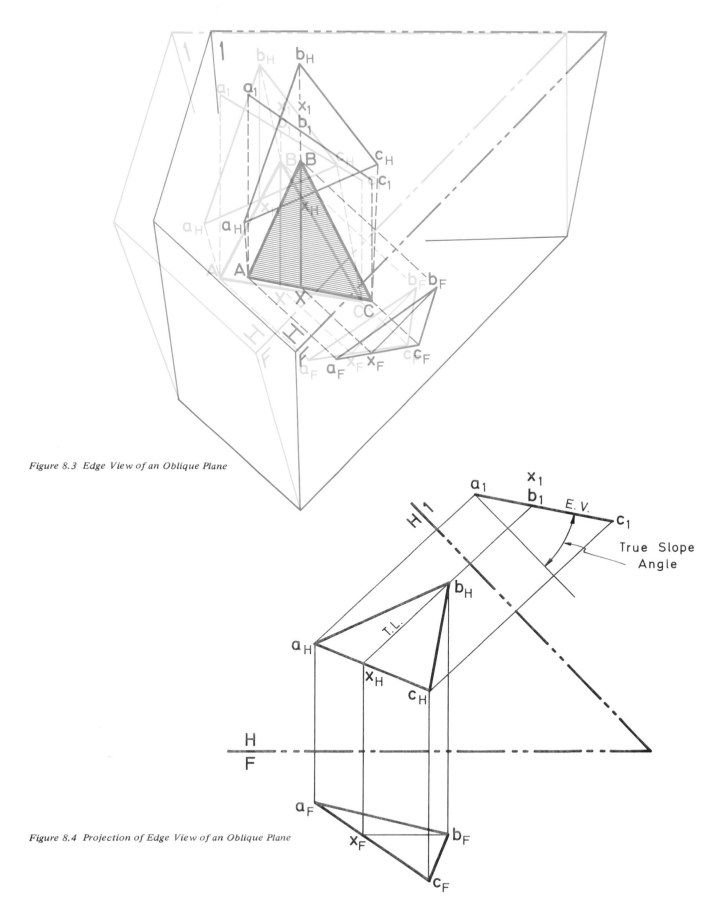

Figure 8.3 Edge View of an Oblique Plane

Figure 8.4 Projection of Edge View of an Oblique Plane

8.3 TRUE SHAPE OF AN OBLIQUE PLANE – NORMAL VIEW

The true shape of an oblique plane cannot be obtained from the horizontal and frontal projections. The true shape is only obtained when the projection plane is parallel to the surface of the oblique plane.

In order to obtain a view of the true shape of an oblique plane an edge view must first be obtained. If we look in a direction perpendicular to the edge of the plane we will see the true shape.

The oblique plane ABC is shown in Figure 8.5. The true length of the line CX is used to obtain the edge view in auxiliary plane 1. Auxiliary plane 2 being perpendicular to auxiliary plane 1 will provide the required true shape. The orthographic projections are shown in Figure 8.6.

The true shape of an oblique plane may also be obtained by rotation as explained in Chapter 11.

Example 8.2 *True Shape*
Given the horizontal and frontal projections of the part shown below, find the area of the inclined surface ABCDEF.

Solution The area must be measured on a true-shape view.

Find the true length of a line on the surface. Line AF is a true length in the horizontal projection. Take a view along this true length to find an edge view in plane 1.

Next obtain a view perpendicular to the edge view. Such a projection is shown in plane 2 and provides a true-shape view of the surface $a_2 b_2 c_2 d_2 e_2 f_2$. The area must be measured on this true-shape view.

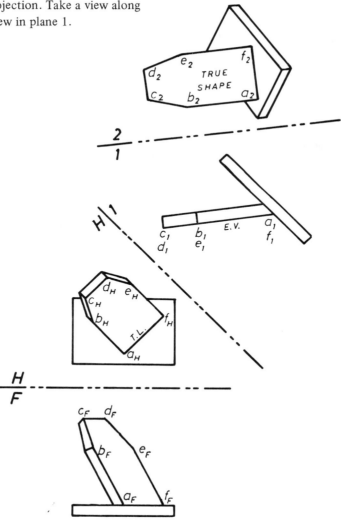

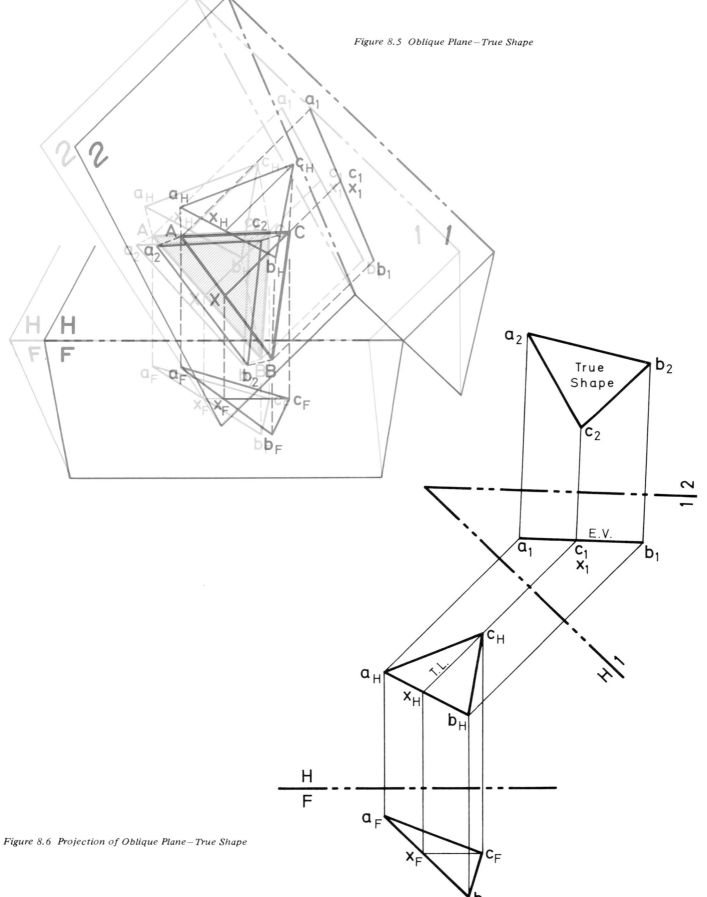

Figure 8.5 Oblique Plane—True Shape

Figure 8.6 Projection of Oblique Plane—True Shape

8.4 DIHEDRAL ANGLE

The angle between two intersecting planes is known as the *dihedral angle.* This angle must be measured in a plane perpendicular to the line of intersection of the two planes in order to obtain its true magnitude.

In order to measure the dihedral angle we must obtain a view which shows both the two planes in an edge view.

If the line of intersection of the two planes is given, a point view of this line will provide the two required edge views of the two planes.

For example in Figure 8.7 the two oblique planes ABC and ADC have the common line of intersection AC. The true length of AC is obtained in auxiliary plane 1 and then the point view is found in auxiliary plane 2. Plane 2 will provide the required edge views on which the dihedral angle can be measured. The orthographic views are shown in Figure 8.8.

The dihedral angle may also be obtained by rotation or as explained in Chapter 11.

Example 8.3 *Dihedral Angle*
Find the angle between the planes ABC and DBC on the given tetrahedron ABCD.

Solution The line of intersection of the two planes is BC. We must obtain a true length of this line and then take a view perpendicular to the true length.

Auxiliary plane 1 is drawn parallel to $b_H c_H$ to provide the true length $b_1 c_1$. Next auxiliary plane 2 is constructed perpendicular to the true length to provide the edge views of the two planes. The dihedral angle is measured in this view.

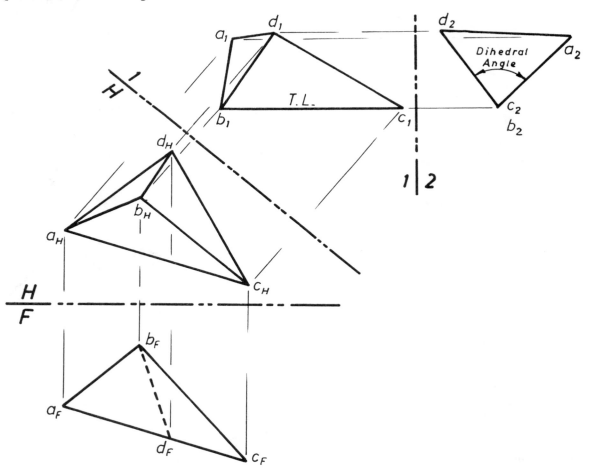

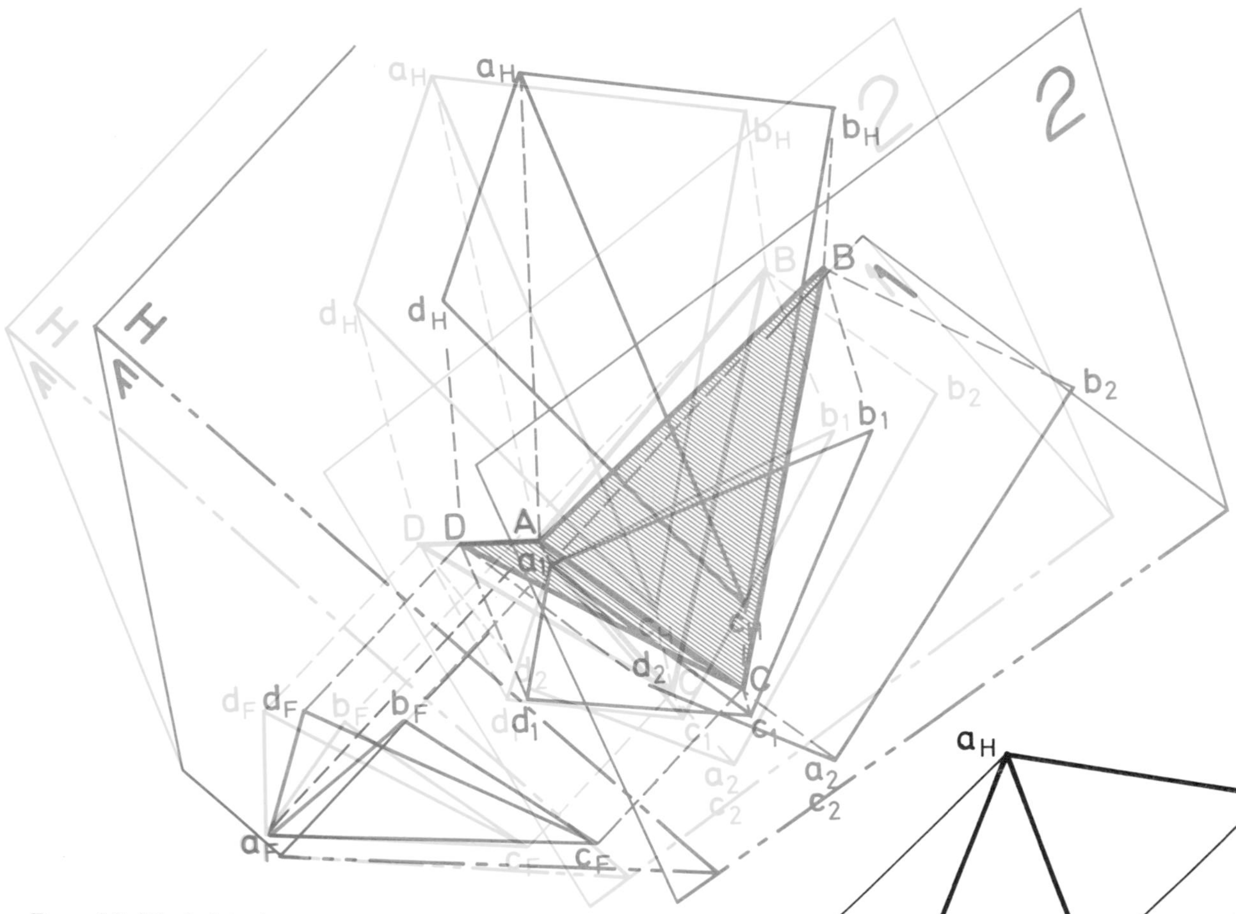

74

Figure 8.7 Dihedral Angle

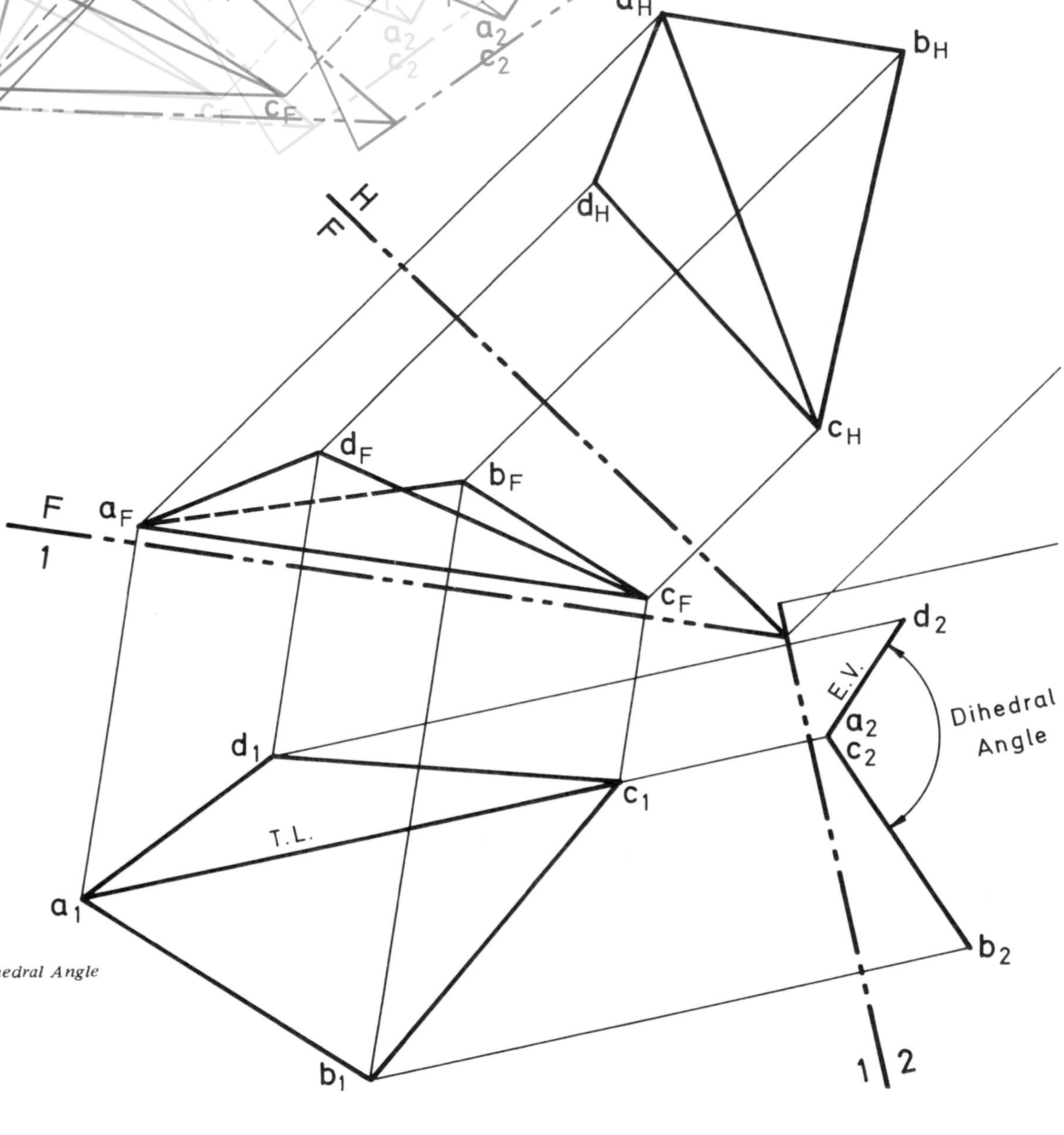

Figure 8.8 Projection of Dihedral Angle

8.5 TRUE ANGLE BETWEEN A LINE AND AN OBLIQUE PLANE

The true angle between a line and an oblique plane can be seen only in a view in which the line appears as a true length *and* the plane appears as an edge. This is illustrated in Figure 8.9. The line is AB and the plane is DEFG.

The procedure for determining the true angle is to first obtain a true-shape view of the plane. Next take an edge view of the given plane in a projection plane parallel to the line. The resulting view will show the line in true length and the plane in an edge view. The true angle can only be measured in this view.

The true angle may also be obtained by rotation as shown in Chapter 11.

Example 8.4 *True Angle between a Line and an Oblique Plane*

Given the line AB and the plane DEFG find the true angle between them.

Solution First obtain a true-shape view of the plane DEFG. Since the line $d_F f_F$ is horizontal therefore it will show as a true length in the horizontal projection $d_H f_H$. A view along this line will provide an edge view of the plane as shown in auxiliary plane 1. A view at right angles to the edge view will show the plane DEFG in true shape in plane 2. Then an auxiliary plane 3 is chosen to be parallel to the line $a_2 b_2$ so that it will be in true length in this view, $a_3 b_3$. The true angle is measured in this view, which shows the line in true length and the plane in edge view.

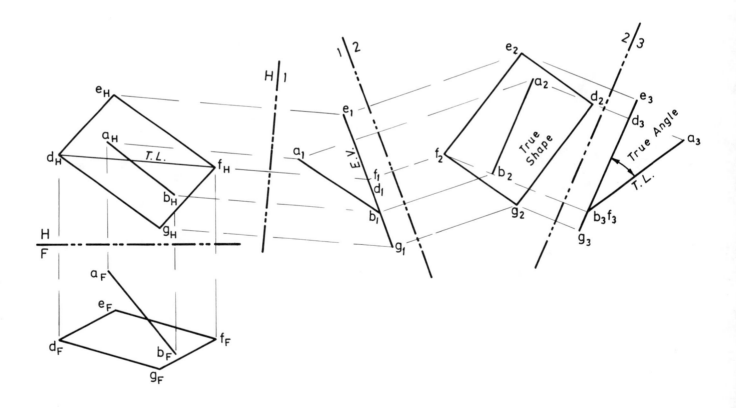

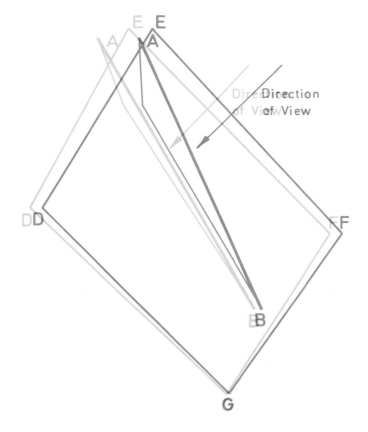

Figure 8.9 Angle between a Line and an Oblique Plane

8.6 A LINE PERPENDICULAR TO A PLANE

A line which is perpendicular to a plane is perpendicular to every line in that plane which goes through the intersection of the given line and the plane. In Figure 8.10 line EF is perpendicular to plane ABCD at E. Every line in the plane which passes through E is also perpendicular to the line EF.

This fact may be more apparent if we examine a view which shows the plane as an edge. We can obtain an edge view of the plane ABCD by taking a view along the true length of the line BC. This is illustrated in Figure 8.11. The plane 1 is chosen so as to be perpendicular to the true length $b_H c_H$ and therefore shows the plane in an edge view. It can now be seen that any line lying in the plane and passing through the point e_1 is perpendicular to $e_1 f_1$.

Example 8.5 *To Draw a Plane Perpendicular to a Given Line*
Given the line EF and a point G, construct a plane through G perpendicular to the line EF.

Solution Obtain a true length of line EF in auxiliary plane 1. The required plane will be in edge view and at right angles to EF. All lines lying in the required plane will be perpendicular to $e_1 f_1$. Therefore draw $g_1 k_1$ perpendicular to $e_1 f_1$ to any convenient point k_1. Since a plane may be represented by three points take a point m_1 at any position on $g_1 k_1$.

Project k_1 and m_1 back to the horizontal plane selecting any point on the projecting lines to locate k_H and m_H. Project down to the frontal plane to locate k_F and m_F by the usual methods.

Plane GKM is the required plane at right angles to line EF.

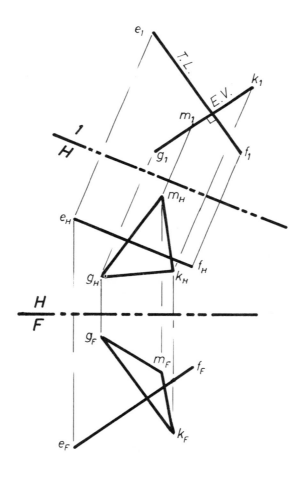

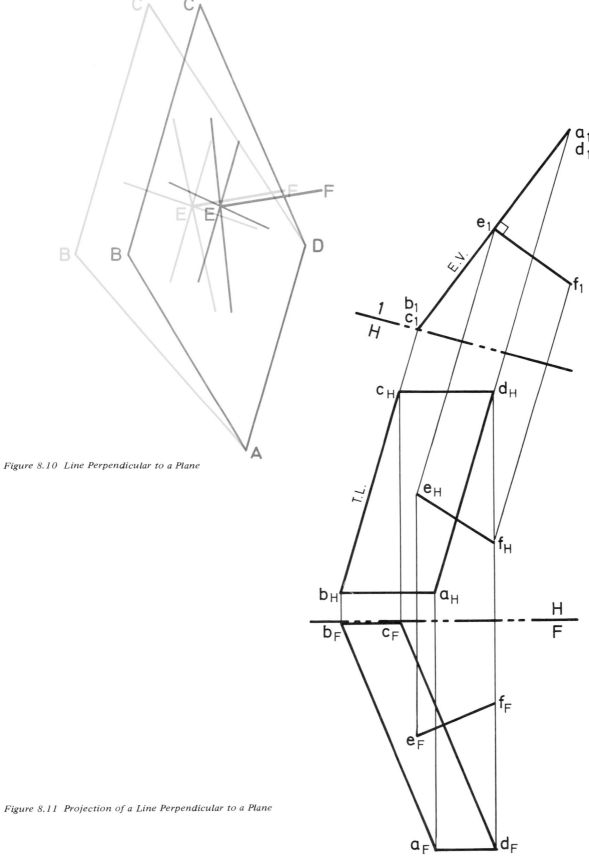

Figure 8.10 Line Perpendicular to a Plane

Figure 8.11 Projection of a Line Perpendicular to a Plane

8.7 VISIBILITY OF POINTS AND LINES

In orthographic projection the line of sight from a point, call it A, is projected on to a plane of projection. If another point or line or surface lies between the point A and the projection plane, then A is not visible on that plane. The situation is illustrated in Figure 8.12 which shows a line AB and a plane surface in space. Note that in the horizontal projection we can see the ends A and B of the line. In the frontal projection, however, the line is hidden by the intervening plane surface. Such hidden lines are usually indicated by a dashed line as shown in the orthographic projection of Figure 8.13.

8.8 VISIBILITY OF NON-INTERSECTING LINES

In the case of non-intersecting or skew lines it is often necessary to determine which line is nearer to the projection plane. In many cases this can be done simply by inspection of the views and noting which points are obviously visible in the projection. However for more complicated situations we must use a more detailed analysis. The principle used can then be applied to determine the visibility of lines in planes or solid objects.

Consider the two skew lines CD and EF in Figure 8.14 and their orthographic projections in Figure 8.15. In the horizontal projection the lines *appear* to intersect. There are actually two points represented at this crossing point, one point on EF and one on CD. Let these points be s_H and t_H. These points will show up on the frontal view as the two separate points s_F and t_F. Similarly the crossing point in the frontal view is actually two points, u_F and v_F, as may be seen in the horizontal view.

In the frontal orthographic view of Figure 8.15 we can see that s_F is above t_F, therefore we may say that line EF is above CD at this point. Similarly in the horizontal projection u_H is in front of v_H. This means that line CD is in front of line EF at that point. We can say that point T is not visible in the H projection and that point V is not visible in the F projection.

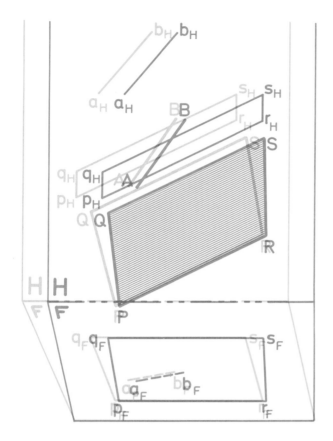

Figure 8.12 Visibility–General Situation

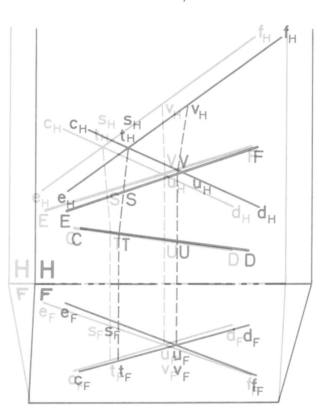

Figure 8.14 Visibility–Non-Intersecting Lines

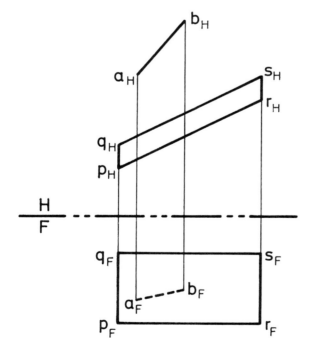

Figure 8.13 Projection of Visibility–General Situation

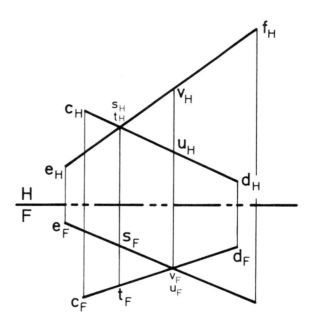

Figure 8.15 Projection of Visibility–Non-Intersecting Lines

8.9 VISIBILITY OF LINES ON SOLID OBJECTS—
Example 1

As an example of visibility consider the tetrahedron shown in Figure 8.16. In the analglyph model the visibility of all the edges is apparent. However when working with orthographic views the double point method may be used to determine the visibility of any line.

To start we may use the fact that all the outer boundaries of all solid objects will be visible in all views. Therefore we may draw in solid lines the boundaries $a_H b_H$, $b_H d_H$, and $a_H d_H$. In the frontal view also the lines $a_F b_F$, $b_F c_F$, $c_F d_F$, and $d_F a_F$ will be solid.

Now determine the visibility of the lines $a_F c_F$ and $d_F b_F$ in the frontal view. Consider the intersection of these projections at the double point x_F on $a_F c_F$ and y_F on $d_F b_F$. Project this point up to the horizontal view to locate the points x_H and y_H. In the horizontal view the point x_H is in front of the point y_H. Therefore y_F is not visible in the frontal view. That is, the line $d_F b_F$ is hidden and should be shown dotted. By the same reasoning the line $a_F c_F$ is therefore shown solid.

To determine the visibility in the horizontal view, consider the frontal view. Here we see that $c_F d_F$ is above $a_F d_F$. Therefore $c_H d_H$ is visible. Similarly $c_H a_H$ and $c_H b_H$ are visible and so are shown as solid lines. In general terms, to determine the visibility of a line or a point in one view, its position in the adjacent view must be examined.

When you are familar with the concept of visibility the procedure can be simplified by considering the points c and y only in the projection Figure 8.17. To determine the visibility in the H view look at the F view and note that c_F is above y_F. Therefore C is visible in the H view and $c_H d_H$, $c_H b_H$, and $c_H a_H$ are drawn in solid lines.

Now look at the H view to determine the visibility in the F view and note that c_H is in front of y_H, therefore y_H is not seen and thus $b_F d_F$ is shown dotted.

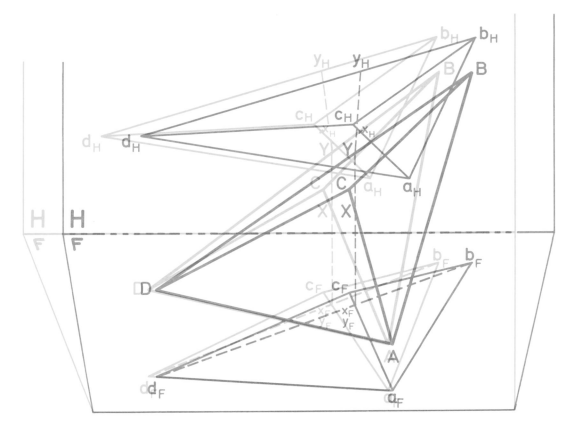

Figure 8.16 Visibility of a Tetrahedron

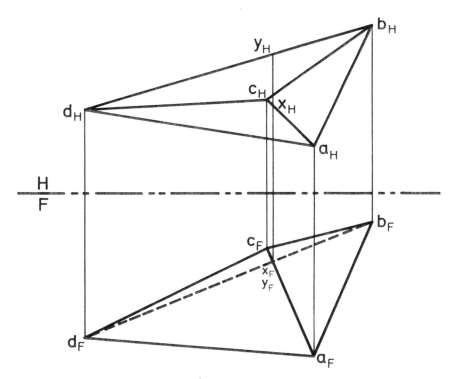

Figure 8.17 Projection of Visibility of a Tetrahedron

8.10 VISIBILITY OF LINES ON SOLID OBJECTS — Example 2

Two orthographic projections of a solid object are shown in Figure 8.18. In Figure 8.18(a) the object is shown with the corner points joined by solid lines. This, of course, is incorrect as some of the lines will not be visible and should be shown dotted.

First use the principle that all the outer boundaries of all solid objects will be visible in all views. Therefore we can immediately outline these edges as shown in Figure 8.18(b).

In order to determine the visibility of the interior lines, apply the principle of double points. Consider first the points E and A so as to make a decision about the visibility of $e_H a_H$. The point e_H may be considered to be the projection of two points, e_H on EA and x_H on FK. Project down from this double point on the H plane to find x_F on $f_F k_F$. In the frontal projection we note that x_F is above e_F, therefore e_H is not visible in the H view.

Now consider point a_H which is the projection of two points a_H on EA and y_H on BC. Project down to the F plane to locate y_F on $b_F c_F$. It can be seen that y_F is above a_F, therefore a_H is not visible in the H view. Since neither end is visible, then we can draw $e_H a_H$ as a hidden dashed line as shown in Figure 8.18(c).

In a similar manner by considering points on the lines EG, GC, GH, FB, KJ, and EJ, the visibility of the remaining interior lines can be determined. The resulting views are shown in Figure 8.18(c).

As an alternative method consider the interior points b_F and j_F in Figure 8.18(b). In the H view both j_H and b_H are exterior points and therefore the edges on which these points lie are drawn solid. As always we must look at the H view to determine the visibility in the F view. Looking at the H projection we see that j_H is in front of b_H. Therefore j_F is visible and $j_F e_F$ and $j_F k_F$ are shown as solid lines. Since b_H is behind j_H the edges to b_F are shown dotted as indicated in Figure 8.18(c).

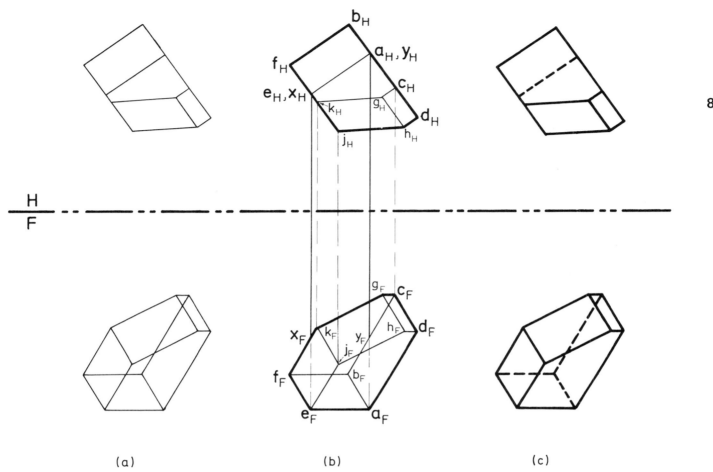

Figure 8.18 Visibility of Lines on a Solid Object

PROBLEMS
Dimensions in millimetres unless otherwise specified.

P 8.1 Determine whether the point D lies in the plane of triangle ABC.

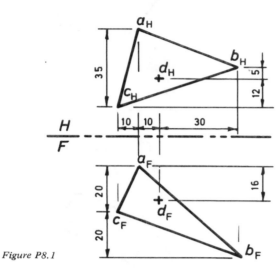

Figure P8.1

P 8.2 Find the true slope angle of plane PQR.

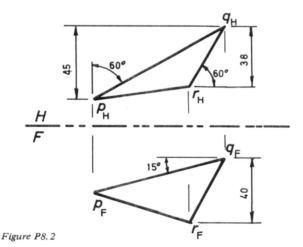

Figure P8.2

P 8.3 As shown in plan view in Figure P 8.3 a sewer line slopes downward at 12% from Main Street to meet another sewer under High Street which has a slope of −30%. What is the true angle between these two lines? Another sewer sloping at −20% under Broadway is joined to the High Street sewer. What is the angle between these two lines?

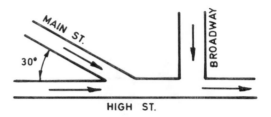

Figure P8.3

P 8.4 Find the true shape of the surface ABC, Figure P 8.4.

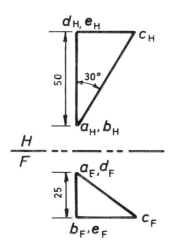

Figure P8.4

P 8.5 The inclined or double auxiliary view 2 shows a circle in true shape. Project back to the first auxiliary view and to the H and F views. What is the true slope of the plane in which the circle lies.

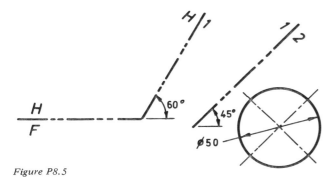

Figure P8.5

P. 8.6 Obtain a true-shape view of face ABC of the tetrahedron ABCD. Project from top view.

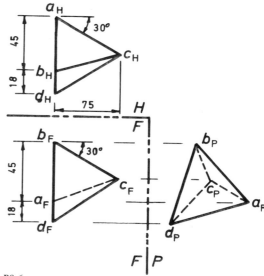

Figure P8.6

P 8.7 The horizontal and frontal projections of a bridge pier are shown in Figure P 8.7. Find the dihedral angle between the faces A and B. Choose a suitable scale for your drawing.

Dimensions in metres

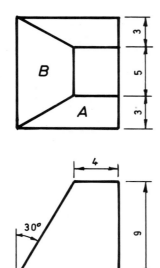

Figure P8.7

P 8.9 A vein of coal is represented by the plane ABC. A mine tunnel PQ is being driven towards the vein. At what angle will the tunnel meet the vein? How much farther must the tunnel be driven in order to meet the vein?

Dimensions in metres

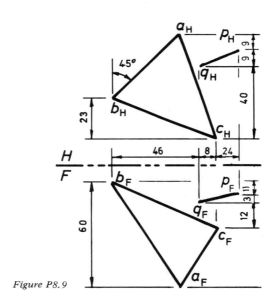

Figure P8.9

P 8.8 Find the dihedral angle between the face A and the bottom of the block in Figure P 8.8.

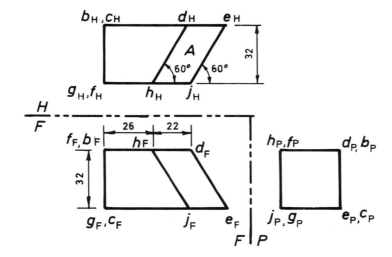

Figure P8.8

P 8.10 The H and F projections of a solid object are shown in Figure P 8.10. Draw the views as projected on to planes 1 and 2. Show correct visibility.

P 8.12 The centre lines of three 6 mm dia rods are shown. Complete the H and F views. Show the correct visibility, omitting hidden lines. The ends of the rods may be sketched by freehand. Construct a profile view.

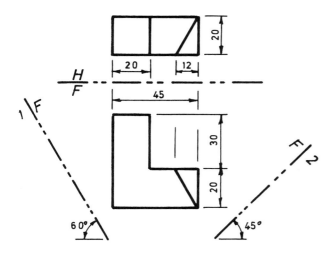

Figure P8.10

P 8.11 Construct a plane ABC perpendicular to line PQ. Let C be the point where PQ pierces the plane. Note that b_H only is given.

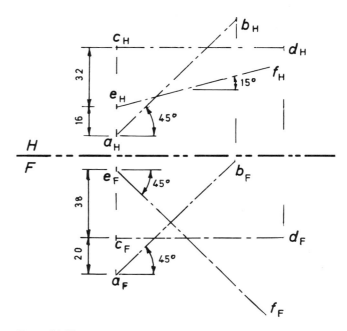

Figure P8.12

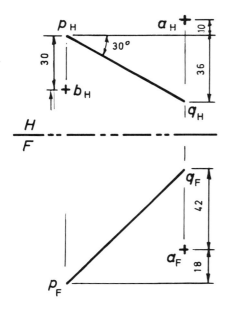

Figure P8.11

9. Distance Between Points, Lines, and Planes

9.1 PERPENDICULAR DISTANCE FROM A POINT TO A LINE

In order to determine the perpendicular distance from a point to a line it is necessary to construct a line from the given point perpendicular to the given line.

Two methods are available to perform the required construction:
1) the line method, and
2) the plane method which uses the plane formed by the given point and the line.

Line Method

In this method we obtain a true-length view of the given line and then drop a perpendicular on to it from the point. This is the required perpendicular line from the point to the line but as constructed it is not itself a true length. Another view is required in order to determine the actual true length of the perpendicular distance.

The situation is illustrated in Figure 9.1 and the procedure may be followed from the orthographic projections in Figure 9.2.

First we find the true length of the given line AB in plane 1. Then draw a perpendicular from c_1 on to the true length $a_1 b_1$, meeting $a_1 b_1$ at d_1. Next the true length of the perpendicular line CD is obtained by establishing the double auxiliary plane 2 parallel to $c_1 d_1$. The true length may be measured as $c_2 d_2$ in plane 2. The true length of CD could also have been obtained by the rotation method.

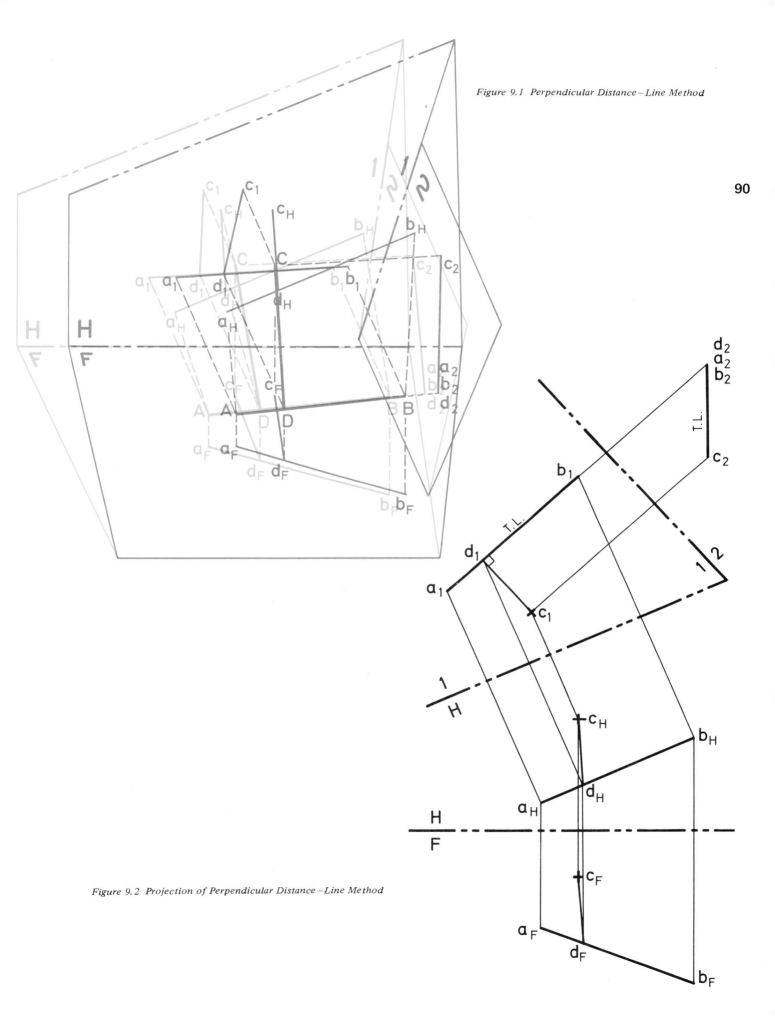

Figure 9.1 Perpendicular Distance–Line Method

Figure 9.2 Projection of Perpendicular Distance–Line Method

90

9.2 PERPENDICULAR DISTANCE FROM A POINT TO A LINE—Plane Method

In this method a plane is formed by the given line and the given point. A true-shape view of this plane is obtained and the perpendicular from the point to the line is constructed on this view. It will be seen in its true length since it is drawn on the true shape of the plane.

The situation is illustrated in Figure 9.3. The procedure may be followed in the orthographic view of Figure 9.4.

Given the line AB and the point C, join CA, CB to form the plane. In order to find the true shape of this plane the procedure outlined in Section 8.3 will be followed.

First obtain a true length of a line in the plane. For convenience the horizontal line CX is drawn because its true length will be shown in the horizontal projection. Next an edge view of the plane is found by looking along the true length as shown in auxiliary plane 1. Finally the true shape is obtained by taking a view at right angles to the edge view as shown in auxiliary plane 2. On this true shape, drop a perpendicular from the point c_2 to meet the line a_2b_2 at d_2. This is the required perpendicular distance in its true length. The location of d_1, d_H, and d_F may be obtained by projecting back to these respective planes.

Example 9.1 *Steepest Line Joining a Point and a Line*
Find the true length and the slope of the steepest pipe to connect the point C with the pipe AB.

Solution Form a plane consisting of the point C and the pipe AB, plane ABC. The steepest line joining the point and the pipe will be the steepest line in this plane. Also the steepest line will be perpendicular to any horizontal line in the plane.

Join C to AB in each view. Draw a horizontal line $b_F x_F$ and project to locate x_H. The required steep pipe $c_H d_H$ must be perpendicular to the true length $b_H x_H$. Draw $c_H d_H$ perpendicular to $b_H x_H$. Project to the frontal view to find d_F. The line CD is the required steepest line.

To find the slope and true length draw the auxiliary projection parallel to $c_H d_H$ to find the true length $c_1 d_1$. This view also shows the plane in edge view so that the slope may be found = θ_H as shown.

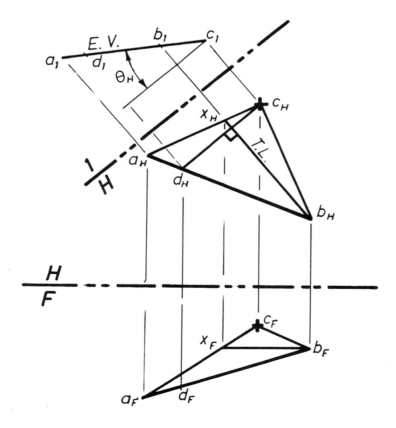

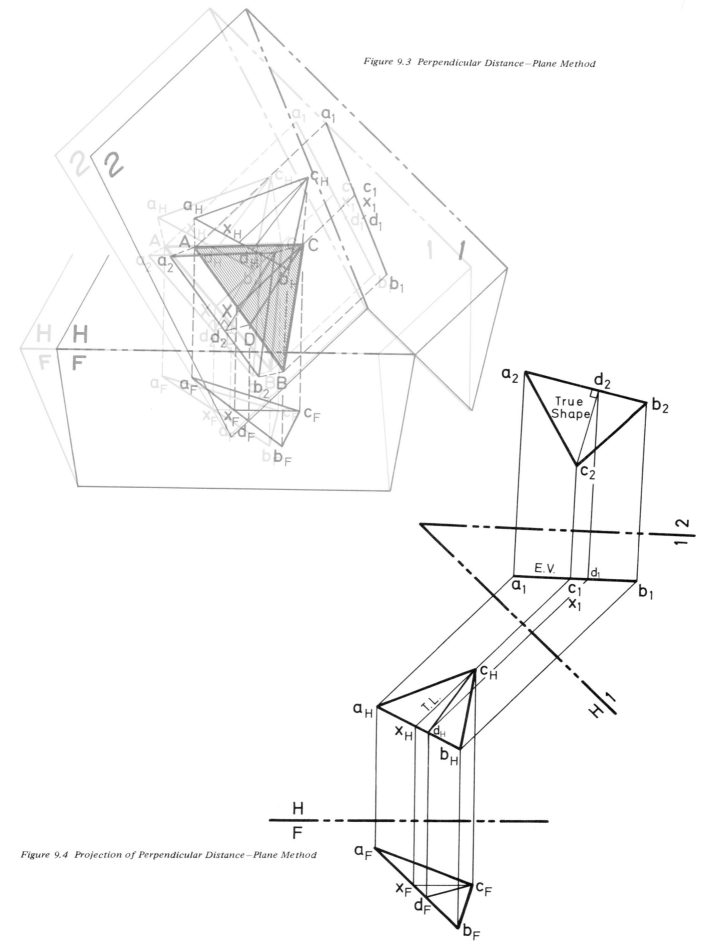

Figure 9.3 Perpendicular Distance — Plane Method

Figure 9.4 Projection of Perpendicular Distance — Plane Method

9.3 PERPENDICULAR DISTANCE BETWEEN TWO SKEW LINES

The shortest distance between two skew lines is a line that is perpendicular to both the two lines. This distance can be obtained by two different methods, a line method and a plane method.

Line Method

In this method a view is obtained showing one of the lines as a point. A perpendicular dropped from this point to the other line will be the shortest distance in its true length.

The situation is illustrated in Figure 9.5. The orthographic projections are shown in Figure 9.6.

The two given skew lines are AB and CD. Select AB and find its true length in the auxiliary plane 1. The view along this line as shown in auxiliary plane 2 will provide a point view of the line a_2b_2. Drop a perpendicular from this point to meet the line c_2d_2 in x_2. The line $a_2b_2x_2$ is the required perpendicular. The points x_1x_H and x_F may be obtained by projecting back to their respective planes. The location of the projection y_1 is found by dropping perpendicular from x_1 to the line a_1b_1. After y_1 is found y_H and y_F are obtained by projecting back to the H and F planes. The required true length is y_2x_2.

Example 9.2 *Angle between Two Skew Lines*

Given the two non-intersecting lines AB and CD find the true angle between them.

Solution The true angle can only be measured in a view which shows both the lines in true length.

Obtain the true length a_1b_1 in auxiliary plane 1. Auxiliary plane 2 will provide a point view of the line a_2b_2. Finally auxiliary plane 3 will show both lines in true length. The true angle may be measured in auxiliary view 3.

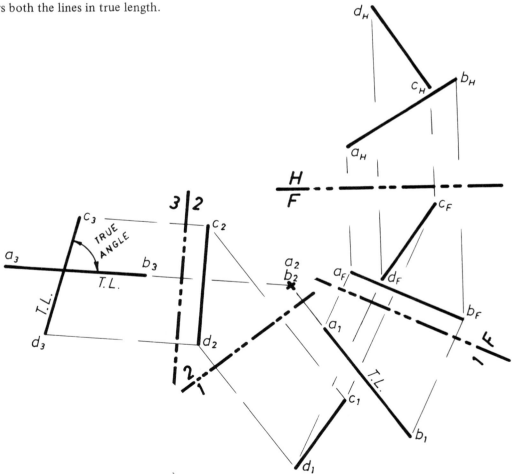

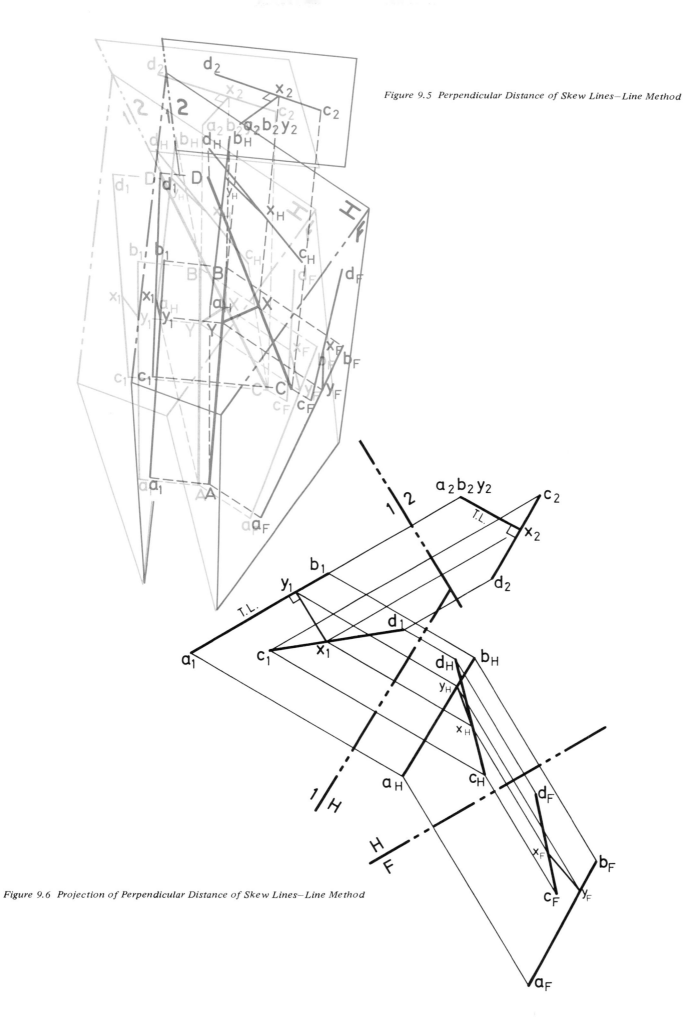

Figure 9.5 Perpendicular Distance of Skew Lines—Line Method

Figure 9.6 Projection of Perpendicular Distance of Skew Lines—Line Method

9.4 PERPENDICULAR DISTANCE BETWEEN TWO SKEW LINES—Plane Method

In this method a plane is formed containing one of the given lines and parallel to the other given line. An edge view of this plane will provide the required perpendicular distance but the exact location of the perpendicular cannot be found in this view. Another view is required showing the common perpendicular line as a point.

The situation is illustrated in Figure 9.7 with the orthographic view in Figure 9.8.

Procedure

Given the horizontal and frontal projections of two skew lines AB and CD.

We will form a plane containing CD and parallel to AB. To do this choose a convenient point on CD, say the end D and draw a line DE parallel to AB. That is through d_H draw $d_H e_H$ parallel to $a_H b_H$ and through d_F draw $d_F e_F$ parallel to $a_F b_F$.

Since an edge view of the plane CDE is required, we will need a view along the true length of a line in this plane. For convenience draw the line $c_F m_F$ parallel to the H plane so that its true length will appear as $c_H m_H$.

A view in the direction of the true length $c_H m_H$ will provide an edge view of the plane CDE. Therefore choose the projection plane 1, perpendicular to $c_H m_H$ as shown. The projection in this plane provides the edge view. The required perpendicular distance is the distance in this view between $a_1 b_1$ and $c_1 d_1$, that is, $r_1 s_1$, and it appears here in its true length.

To locate the precise position of the perpendicular line we form a view in the plane 2 chosen so that it will provide a point view of the line. Plane 2 is therefore perpendicular to the required line and the line $r_2 s_2$ appears as the point of intersection of $a_2 b_2$ and $c_2 d_2$. The points $r_2 s_2$ can be projected back to the other views.

9.4(a) ANGLE BETWEEN TWO SKEW LINES

The angle between two skew lines will appear in its true size in any plane that is parallel to both lines. This is illustrated in the above construction where plane 2 is parallel to both AB and CD. The true angle will be angle $a_2 r_2 c_2$.

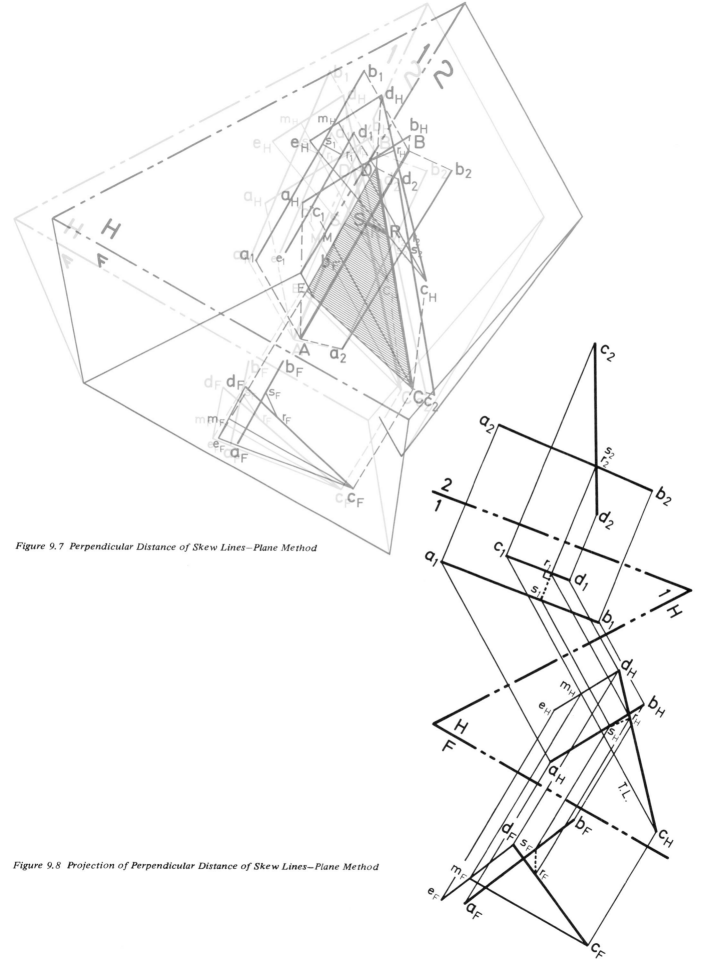

Figure 9.7 Perpendicular Distance of Skew Lines—Plane Method

Figure 9.8 Projection of Perpendicular Distance of Skew Lines—Plane Method

9.5 SHORTEST HORIZONTAL DISTANCE BETWEEN TWO SKEW LINES

This situation is very similar to the previous section. A plane is formed containing one of the lines and parallel to the other as before. The edge view of this plane is formed, again as before, but in this case a horizontal line is drawn instead of a perpendicular line.

An example is shown in Figure 9.9 with the orthographic projections in Figure 9.10.

Procedure

Given the horizontal and frontal projections of two skew lines AB and CD, as shown in Figure 9.9.

We will form a plane containing CD and parallel to AB. To do this choose a convenient point on CD, say the end D and draw a line DE parallel to AB. That is, through d_H draw $d_H e_H$ parallel to $a_H b_H$ and through d_F draw $d_F e_F$ parallel to $a_F b_F$.

Since an edge view of the plane CDE is required we will need a view along the true length of a line in this plane. For convenience draw the line $c_F m_F$ parallel to the H plane so that its true length will appear as $c_H m_H$.

A view in the direction of the true length $c_H m_H$ will provide an edge view of the plane CDE. Therefore choose the projection plane 1, perpendicular to $c_H m_H$ as shown. The projection in this plane provides the edge view. A horizontal line (that is, parallel to the H plane) lying somewhere between $c_1 d_1$ and $a_1 b_1$ in this view will be the required line.

To determine the exact location, a view at right angles to the horizontal will show the horizontal line as a point. Therefore choose a plane 2 as shown. The point of intersection on this plane, of $a_2 b_2$ and $c_2 d_2$ will provide the location of the required line at $v_2 t_2$. These points can now be projected back to the other views.

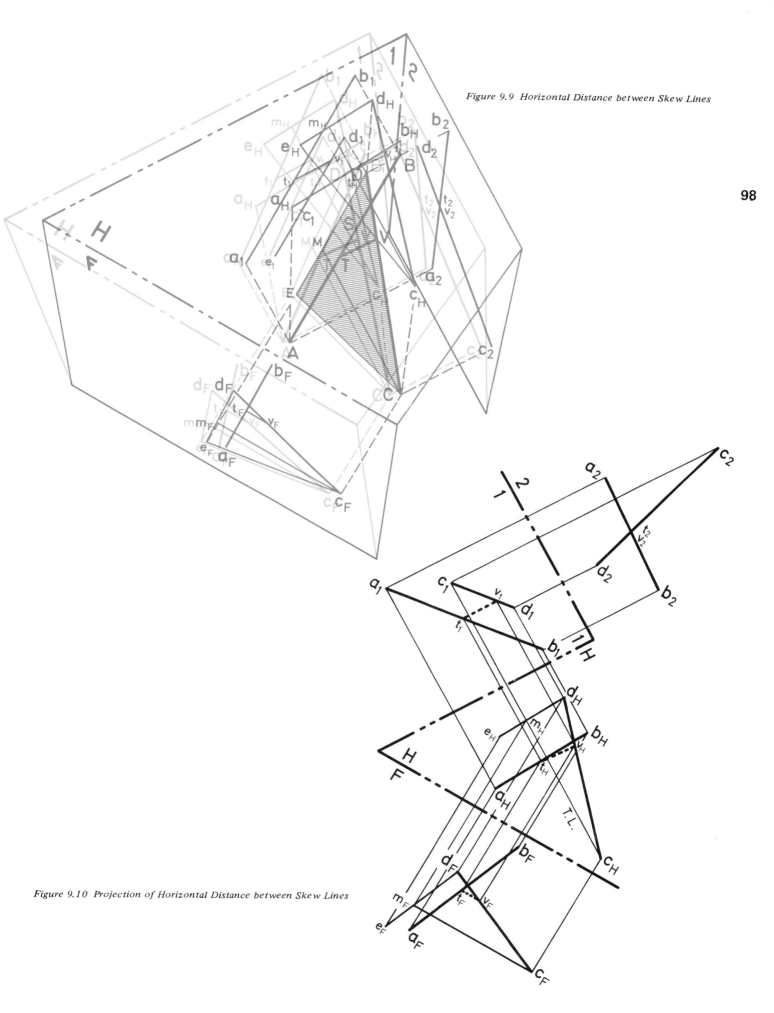

Figure 9.9 Horizontal Distance between Skew Lines

Figure 9.10 Projection of Horizontal Distance between Skew Lines

9.6 PERPENDICULAR DISTANCE FROM A POINT TO A PLANE

Given a point and a plane, the perpendicular distance can be determined in the view showing the plane as an edge. When this view is obtained a perpendicular may be dropped from the point to the plane to determine the position of the intersection of the perpendicular and the plane.

The situation is illustrated in Figure 9.11 and the orthographic projections are shown in Figure 9.12.

Procedure

Give the point P and the plane ABC in the H and F projections as shown.

Obtain a true-length line in the given plane by drawing $b_F m_F$ horizontal to provide the true length $b_H m_H$.

A view along the true length will provide the edge view of plane ABC. Therefore construct the auxiliary plane 1 and obtain the required edge view.

Drop a perpendicular from p_1 to meet $a_1 c_1$ at r_1. This is the required perpendicular distance in its true length.

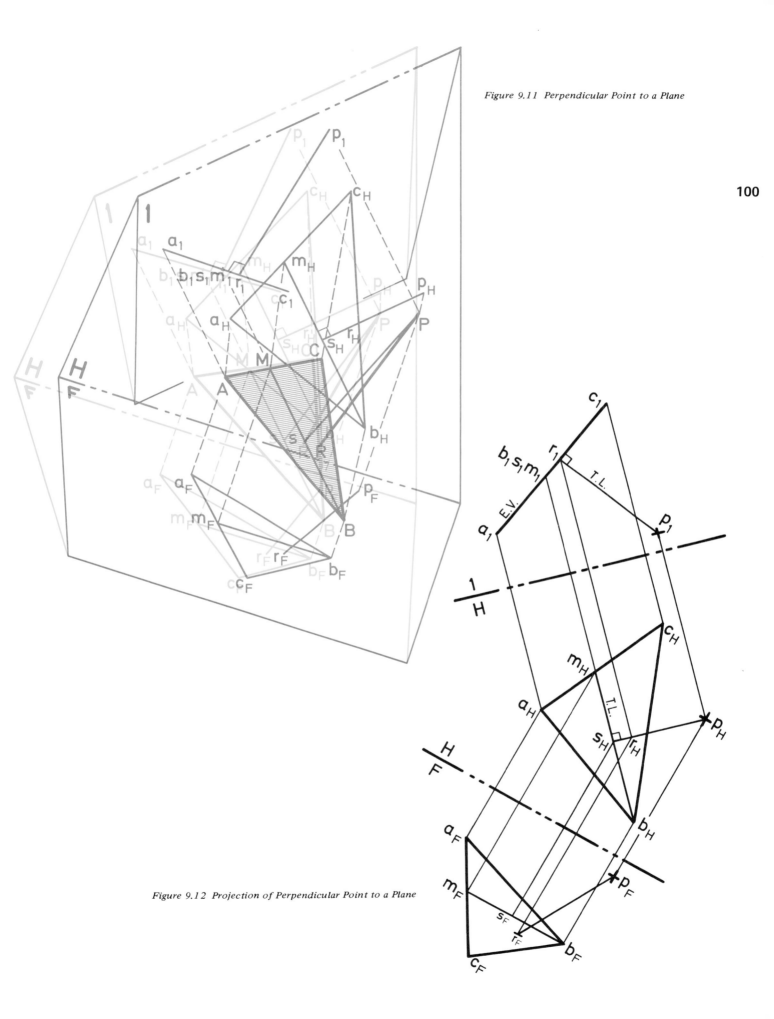

Figure 9.11 Perpendicular Point to a Plane

100

Figure 9.12 Projection of Perpendicular Point to a Plane

PROBLEMS

All dimensions in millimetres unless otherwise noted.

P 9.1 Determine the length, slope, and bearing of the shortest line from point A to the line BC.

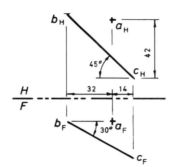

Figure P9.1

P 9.2 The line AB represents the centre line of a buried pipeline. A 90° Tee is to be used to connect a pipe from Y to AB. Find the true length, slope, and bearing of the connecting pipe. How far from B must the Tee be inserted?

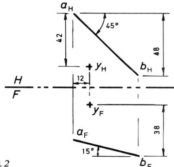

Figure P9.2

P 9.3 Find the shortest distance between the lines PQ and RS.

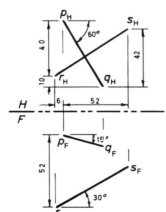

Figure P9.3

P 9.4 Find the shortest horizontal distance between the lines XY and VW.

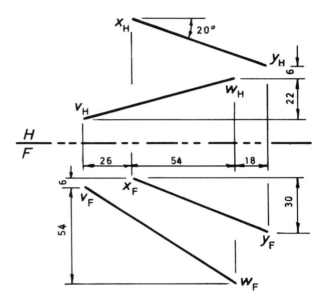

Figure P9.4

P 9.5 Determine the shortest distance on a 20% grade between lines AB and CD.

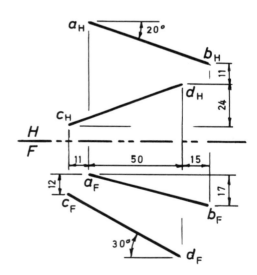

Figure P9.5

P 9.6 Determine whether the wire AB will interfere with the plate PQR. How much clearance exists if the wire clears?

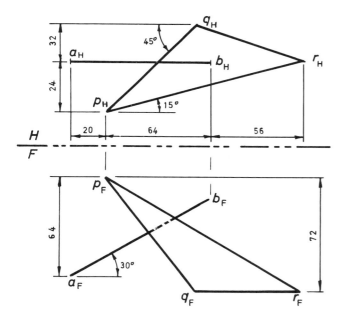

Figure P9.6

P 9.7 Find the shortest distance from the point A to the plane BCD, Figure P 9.7.

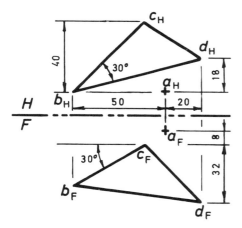

Figure P9.7

P 9.8 A high-voltage line is to be installed from the substation at C to the top of the pole at D. An existing high-voltage line runs from A to B. Find the shortest distance between the two wires AB and CD, Figure P 9.8.

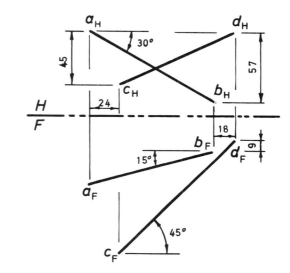

Figure P9.8

P 9.9 A bearing support plate for two shafts is shown in Figure P 9.9. Determine if it is possible to withdraw shaft A to the right with shaft B still in position. That is, is there any interference between the two shafts?

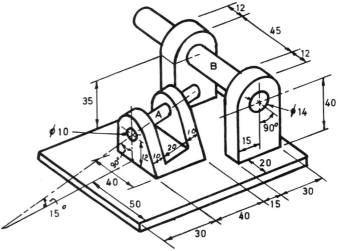

Figure P.9.9

10. Piercing Points

A line will pierce a plane if it does not lie in the plane or is not parallel to the plane. The location of this piercing point is important as it is the basis of Oblique Projection and Perspective Projection. These projections provide useful pictorial views and will be considered in Chapter 12. In addition the drawing of shadows cast by objects is based on the concept of piercing points. Some examples may be found in Problems P 10.5 to P 10.10.

The piercing point of a line with a plane may be found by either the Edge View method or the Cutting Plane method.

10.1 EDGE VIEW METHOD

In this procedure an edge view of the plane and a view of the line is obtained by the usual method. The piercing point will be the point of intersection in this view of the plane with the line. The situation is shown in Figure 10.1 with the orthographic projection in Figure 10.2.

Procedure

Given the horizontal and frontal projections of the plane ABC and the line MN, find the point where MN pierces the plane ABC.

To obtain an edge view of the plane we need to look along the true length of a line in the plane. For convenience draw $b_H k_H$ in Figure 10.2 parallel to the frontal plane to obtain the true length $b_F k_F$. A view in this direction is obtained in plane 1 perpendicular to $b_F k_F$ and shows plane ABC in edge view $a_1 b_1 c_1$.

The point of intersection p_1 is obtained where $m_1 n_1$ intersects $a_1 b_1 c_1$ and this is the required piercing point which is projected back to the other views.

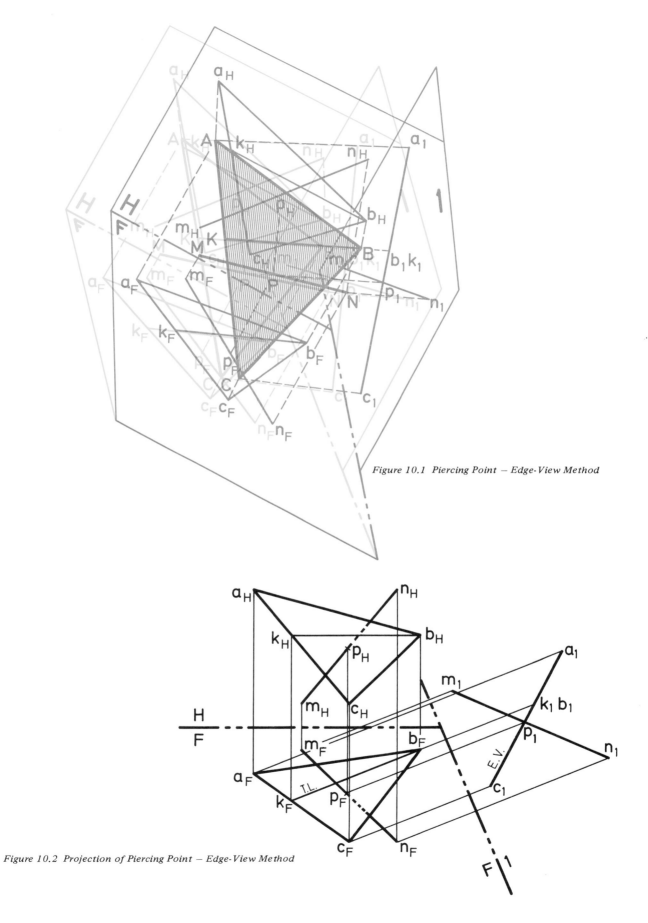

Figure 10.1 Piercing Point — Edge-View Method

Figure 10.2 Projection of Piercing Point — Edge-View Method

10.2 PIERCING POINTS OF LINES WITH PLANES—Cutting Plane Method

In this method a plane is formed containing the given line. This plane may be a vertical plane or a horizontal plane. The *line* of intersection of this plane with the given oblique plane, will intersect the given line or be parallel to it. This is because these two lines (the line of intersection and the given line) are in the same cutting plane. When a vertical cutting plane is used, the piercing point will appear in the frontal projection. Conversely when a horizontal cutting plane is chosen the piercing point will appear in the horizontal projection. The situation is illustrated in Figure 10.3 and Figure 10.4.

Procedure

Given the line MN and the plane ABC we wish to find the piercing point of MN in ABC.

We will use a vertical cutting plane RSVT. Since this is a vertical plane it will appear as a line in the H projection. Choose convenient points r_H and s_H as shown in Figure 10.4 along the line of $m_H n_H$ since MN lies in this plane. This plane will intersect the plane ABC at x_H and y_H.

In the frontal projection let $t_F v_F$ be any convenient distance below the horizontal plane. Locate points x_F and y_F from the horizontal projection where the plane intersects $a_F c_F$ and $a_F b_F$. The intersection of $x_F y_F$ with $m_F n_F$ at p_F is the required piercing point. This point can be projected up to the H plane to locate p_H.

Example 10.1 *Piercing Point Problem*

With A as a point light source, find the shadow cast by the line BC on plane DEFG. Use the cutting-plane method.

Solution   The problem is to find the piercing points of the lines AC and AB on the given plane.

Pass a vertical plane through $a_H c_H$. This will cut the given plane edges $d_H g_H$ at m_H and $d_H e_H$ at p_H. Locate m_F and p_F on the frontal view. Project $a_F c_F$ to meet the line $m_F p_F$ at x_F. Project upward to find x_H. X is one end of the required shadow line.

The end Y of the shadow line is obtained in a similar manner as shown.

XY is the required shadow of BC.

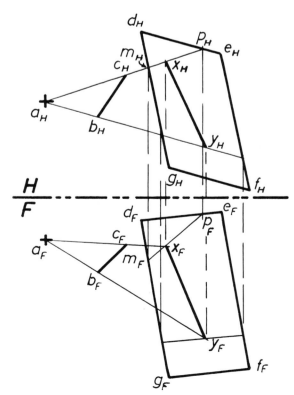

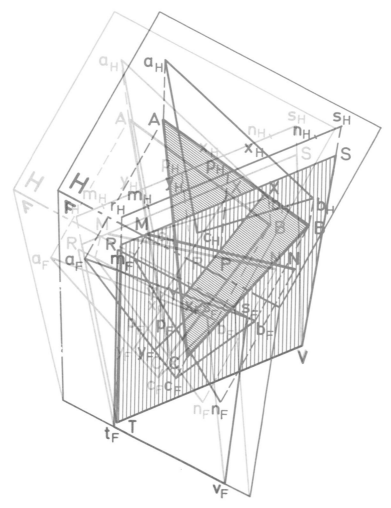

Figure 10.3 Piercing Point—Plane Method

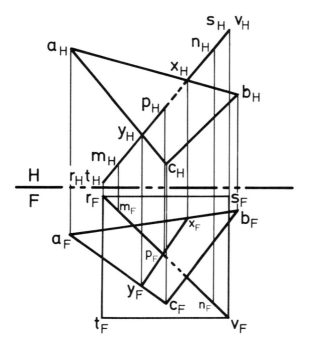

Figure 10.4 Projection of Piercing Point—Plane Method

PROBLEMS

All dimensions in millimetres unless otherwise stated.

P 10.1 Determine the intersection of line AB with plane CDE. Use edge-view method. Show correct visibility.

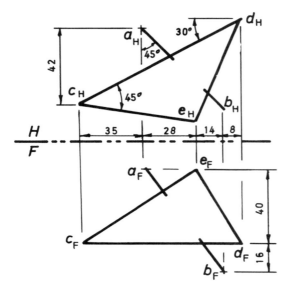

Figure P10.1

P. 10.2 Use the cutting-plane method to find the intersection of line PQ with plane RST.

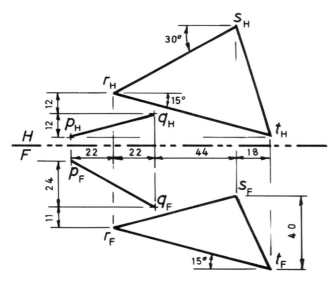

Figure P10.2

P 10.3 An equilateral triangular-base right pyramid is shown in Figure P 10.3. Line AB passes through the pyramid. Find the true length of the portion of AB within the pyramid.

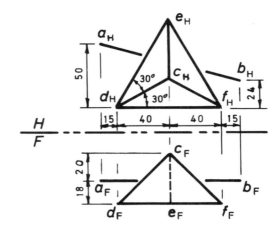

Figure P10.3

P 10.4 In a radiation laboratory a triangular lead plate BCD acts as a shield to protect part of the surface EFGH from gamma rays coming from the point source A. What is the area of the shadowed part of EFGH which is protected from rays by the shield?

$$A = -160, -4, -13$$
$$B = -118, -26, -64$$
$$C = -84, -26, -54$$
$$D = -90, -50, -22$$
$$E = -112, -32, -77$$
$$F = -69, -9, -101$$
$$G = -12, -64, -36$$
$$H = -55, -87, -12$$

Shadows In the following problems the rays of light from the sun are assumed to be parallel and to travel in straight lines.

P 10.5 Shadow cast by a point on a surface. Given the direction of the light rays, find the shadow cast by point A on the surface BCDE. Find the piercing point of a light ray through A and piercing the plane. Use a vertical cutting plane.

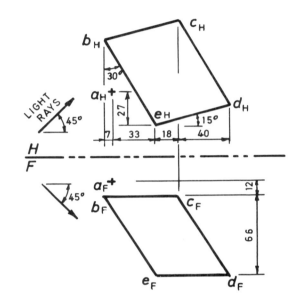

Figure P10.5

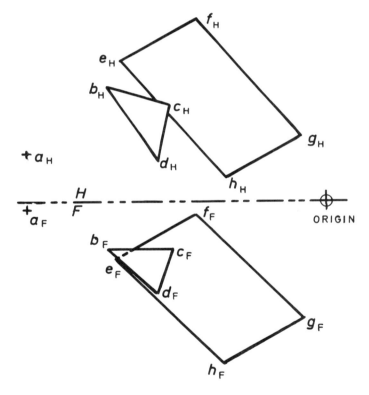

Figure P10.4

P 10.6 Draw the shadow cast by the line MN on the surface RSTU.

P 10.8 Locate the shadow of the point A on the cylinder.

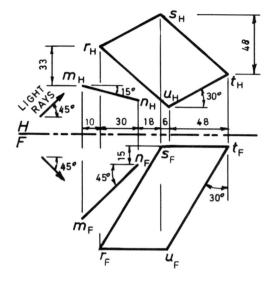

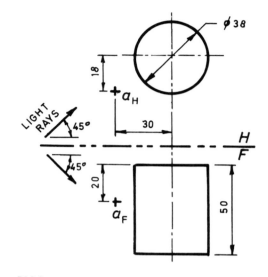

Figure P10.8

Figure P10.6

P 10.7 Draw the shadow of the surface BCD on the surface EFGH.

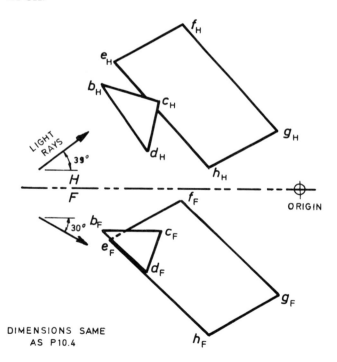

DIMENSIONS SAME AS P10.4

Figure P10.7

P 10.9 Determine the shadow of the line MN on the cylinder.

P 10.10 Determine the shadow cast by the cylinder on the ground.

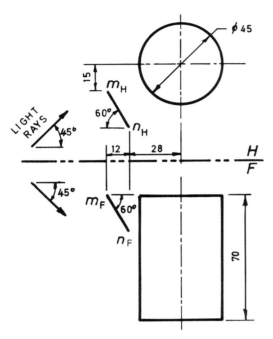

Figure P10.9

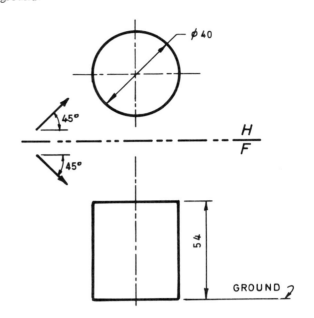

Figure P10.10

11. Rotation

Rotation involves changing the position of a point, line, or surface by revolving it about an axis. The purpose is to obtain a view in a particular direction. In previous chapters we have obtained the view required by changing the position of the observer and developing a projection from the new position. This was done by the use of auxiliary planes while the object remained stationary. When using the rotation method the observer remains stationary and the position of the object changes. An example was described previously (Section 5.4) to find the true length of a line by rotation.

The three basic elements of rotation are:
1) the axis of rotation
2) the centre of rotation
3) the path of rotation.

11.1 ROTATION OF A POINT ABOUT A LINE

We will consider a point A rotating in a circular path about a straight-line axis. The path will lie in a plane which is perpendicular to the axis of revolution. The centre of the circular path lies on the axis.

The elements of rotation are shown in Figure 11.1 with their orthographic projections in Figure 11.2. Note that the path of revolution will appear as an ellipse when foreshortened. The auxiliary plane 1 shows the axis in a point view. When the axis of revolution appears as a point (as in auxiliary plane 1) the path of revolution appears as a circle. Also note that the edge view of the path of revolution is perpendicular to the axis as shown in the horizontal projection. In the frontal view the path is an ellipse. The shape of the ellipse is determined from the auxiliary view. Points on the ellipse are obtained by transfer from the auxiliary view to the frontal view.

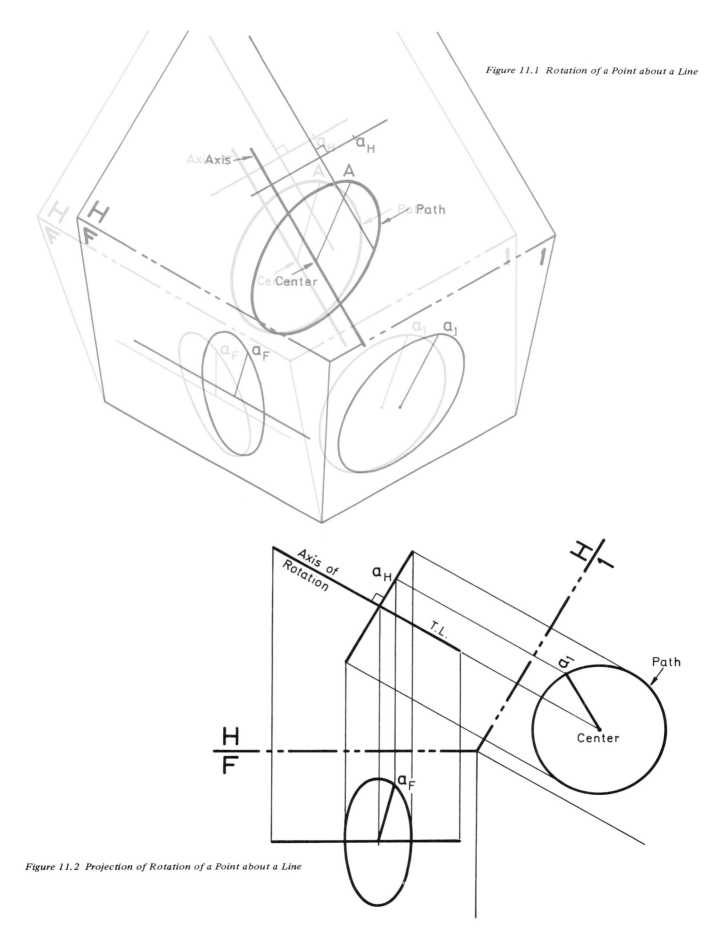

Figure 11.1 Rotation of a Point about a Line

Figure 11.2 Projection of Rotation of a Point about a Line

11.2 ROTATION OF A LINE ABOUT A LINE AXIS

Four general cases may be considered when a line is revolved about another line.

If the two lines are parallel the revolved line will generate a cylinder as shown in Figure 11.3.

When the two lines are skewed to each other, that is not parallel and not intersecting, then a hyperboloid of revolution will result (Figure 11.4). Note that the surface appears curved but is made up of a series of straight lines. This is the basis of some hyperbolic roofs in buildings and the bevel hypoid gear.

A cone will result when the two lines intersect as shown in Figure 11.5. This provides a convenient method of finding the true length and the angle a line makes with the principal planes as explained in Section 5.4.

Finally, if the two lines are perpendicular a circular plane will be the result of revolving one about the other, as shown in Figure 11.6.

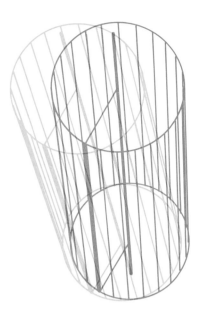

Figure 11.3 Parallel Lines

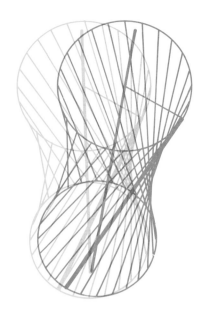

Figure 11.4 Skewed Lines

Figure 11.5 Intersecting Lines

Figure 11.6 Perpendicular Lines

11.3 ANGLE BETWEEN A LINE AND THE PRINCIPAL PLANES

The angle a line makes with a principal plane may be determined by rotating the line until it is parallel to the adjacent principal plane. The axis of rotation is chosen to be perpendicular to the plane to which we require the angle of inclination.

This concept of rotation will be used in this section to determine the angle a line makes with all three reference planes. Either end of the line may be used as the apex of the cone of revolution.

Angle with the Horizontal Plane, θ_H

The angle a line makes with the horizontal plane is obtained by rotating the line about an axis perpendicular to the horizontal plane. The line is rotated until it is parallel to the frontal plane. This is shown in Figure 11.7 where the line AB is rotated about the axis CD. Note that the true length of the line will appear in the frontal view. The orthographic projections are shown in Figure 11.8. The subscript r indicates the rotated position. For example, b_H is rotated to b_{r_H}.

Angle with the Frontal Plane, θ_F

The angle the same line, AB, makes with the frontal plane is shown in Figure 11.9 and Figure 11.10. Note that the axis EF is perpendicular to the frontal plane. Also the line is rotated until it is parallel to the horizontal plane.

Angle with the Profile Plane, θ_P

In this case the line is rotated about axis GH perpendicular to the profile plane. Rotate the line until it is parallel to either the horizontal or the frontal plane. The angle, θ_P, may be measured in the horizontal plane or in the frontal plane as shown in Figure 11.12.

General

The general rule is that to find the angle a line makes with a plane it is rotated about on axis perpendicular to the plane. Rotation is continued until the line is parallel to the adjacent plane.

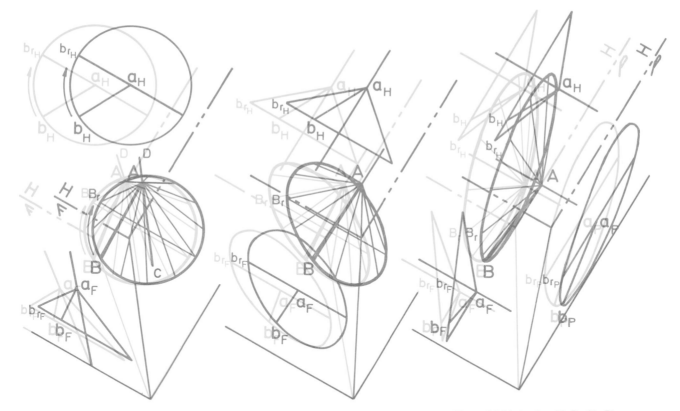

Figure 11.7 Angle with Horizontal Plane

Figure 11.9 Angle with Frontal Plane

Figure 11.11 Angle with Profile Plane

116

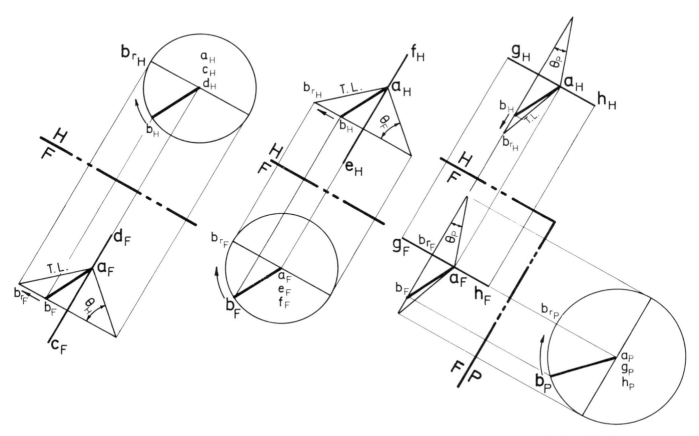

Figure 11.8 Projection of Angle with Horizontal Plane

Figure 11.10 Projection of Angle with Frontal Plane

Figure 11.12 Projection of Angle with Profile Plane

11.4 ANGLE BETWEEN A LINE AND AN OBLIQUE PLANE BY ROTATION

The angle a line makes with an oblique plane will not be seen in the horizontal and frontal projections. In order to determine the angle we must obtain an edge-view and a true-size view of the plane and then rotate the line in these views.

The procedure for finding the true shape of an oblique plane was developed in Section 8.3 and Figure 8.5 on page 72. Therefore the procedure will not be repeated here. The piercing point P is seen in the edge view of the plane.

Given the horizontal and frontal projections of the line DE and the plane ABC obtain an edge view and true-shape view as shown in Figure 11.13.

To obtain the angle between the line DE and the plane ABC, apply the general rule that the line is rotated about an axis perpendicular to the plane. Therefore, rotate the line in the true-shape view (auxiliary view 2) of the plane until it is parallel to auxiliary plane 1. The true length of the line will then be shown in Auxiliary View 1 and the angle θ is measured in this plane. For purposes of illustration only the auxiliary planes 1 and 2 are shown in Figure 11.13. The orthographic projection of the complete problem is shown in Figure 11.14.

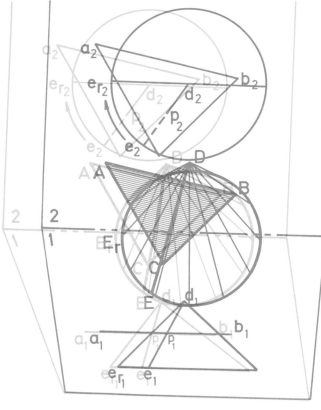

Figure 11.13 Angle between a Line and an Oblique Plane

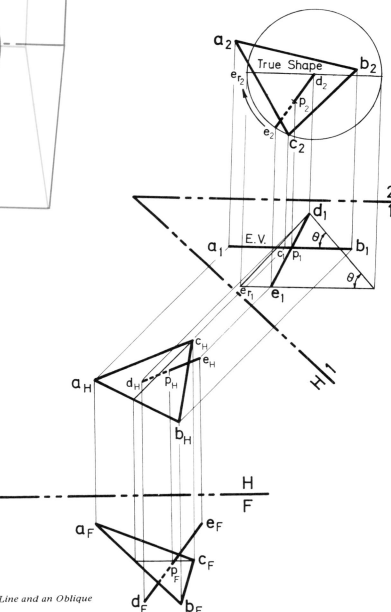

Figure 11.14 Projection of Angle between a Line and an Oblique Plane

11.5 POSITION OF A LINE AT SPECIFIED ANGLES

It is sometimes necessary to determine the location of a line which must make certain specified angles with two of the principal planes. We may call this a problem in synthesis, as opposed to analysis, as we are not given the line but must create, or synthesize it.

This type of problem may be solved by application of the concept of the cone generated by rotating a line.

As an illustration suppose we require a line which makes an angle of 40° with the horizontal plane and 25° with the frontal plane. Also assume that we are given the location of one end, A, of the line.

It is apparent that all the lines which pass through the point A and make an angle of 40° with the horizontal will lie on the surface of a cone. The axis of this cone will be perpendicular to the horizontal plane as shown in Figure 11.15(a). The cone angle will be 40°.

Similarly all the lines passing through A which make an angle of 25° with the frontal plane will lie on a cone. The axis of this cone will be perpendicular to the frontal plane as shown in Figure 11.15(b).

The required line must satisfy both the above conditions. In addition, since one end of the line is at point A the other end must lie on the base of each cone. This means that the sides or elements of the two cones must be the same length. The two cones are shown together in Figure 11.15(c) and it can be seen that there are two solutions, AB and AC.

Actually there are four possible solutions as shown in Figure 11.16(c). These are obtained by extending the cones. There will be no solution if the two angles total more than 90°.

General Procedure

The construction may be all drawn on one drawing as in Figure 11.16(c).

Draw the triangular view of one cone first to any convenient size. This determines the true length of the cone element. Now draw the triangular view of the other cone, making sure that the side of this isosceles triangle is the same length as the side of the first triangle. That is, the true lengths must be the same magnitude. Next draw the circular views for each cone. The point of intersection is on the base of each cone as shown at B and C in Figure 11.16(c). The other two solutions may be obtained by considering the cones to be extended. In this case the bases of the cones will intersect at D and E. The four possible solutions are AB, AC, AD, and AE.

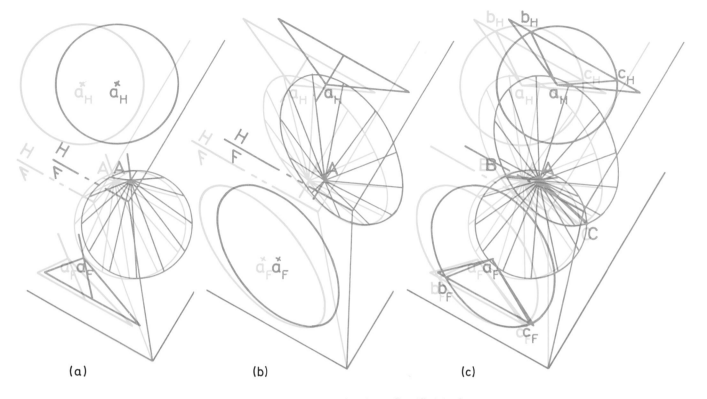

Figure 11.15 Location of a Line at Specified Angles

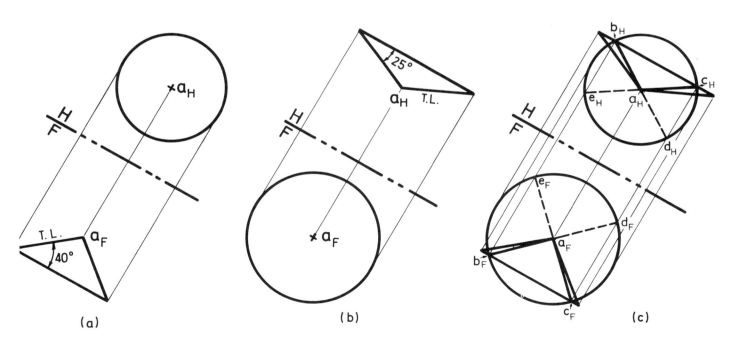

Figure 11.16 Projection of Location of a Line at Specified Angles

11.6 ROTATION OF A PLANE ABOUT A LINE—True Shape of an Oblique Plane

The true shape of an oblique plane was obtained beforehand (Section 8.3) by developing a double auxiliary view. In this section the true shape will be obtained by rotating the plane about a true-length line in the plane. The oblique plane is rotated until it is parallel to the plane containing the true-length view.

As an example consider the plane ABC shown in Figure 11.17 and the projections in Figure 11.18. We will require an edge view of the plane in order to rotate the plane about a line in the plane.

First we must obtain the true length of a line in this plane. For convenience choose the horizontal line $c_F x_F$ and obtain its true length $c_H x_H$ in the horizontal plane. We are going to revolve the plane about this line as an axis. Therefore, we need a point view of the axis CX.

The second step is to obtain this point view of the axis CX by taking a view parallel to the true length $c_H x_H$. This view in auxiliary plane 1 will also provide an edge view of the plane.

The third step is to rotate the plane about the point view of the line $c_1 x_1$. Point a_1 will follow a circular path with centre at $c_1 x_1$. The plane is rotated until it is parallel to the horizontal plane. That is a_1 is rotated to a_{r_1}. Similarly b_1 rotates to b_{r_1}.

Finally obtain the horizontal projection of the rotated plane. Point a_H moves perpendicularly to the axis $c_H x_H$ to point a_{r_H}. Likewise point b_H moves to b_{r_H} during the rotation. The final true shape is shown in the horizontal projection as the plane $a_{r_H} b_{r_H} c_H$.

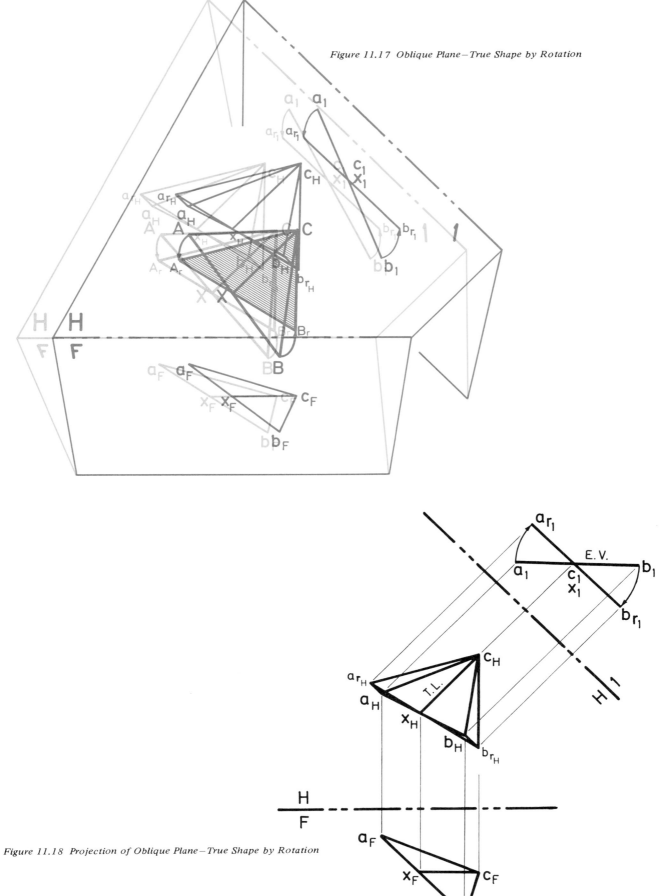

Figure 11.17 Oblique Plane – True Shape by Rotation

Figure 11.18 Projection of Oblique Plane – True Shape by Rotation

11.7 DIHEDRAL ANGLE BY ROTATION

The angle between two planes is known as the dihedral angle. A method of determining this angle was developed previously (Section 8.4) using a double auxiliary view to show the planes in edge view. In this section we will find the dihedral angle order to obtain a true-shape view of it. The dihedral angle will be measured on the true-shape view.

First obtain a true-length view of the line of ADC shown in Figure 11.19 and 11.20. We will form a plane perpendicular to their line of intersection AC. Then we will rotate this plane in order to obtain a true-shape view of it. The dihedral angle will be measured on the true-shape view.

First obtain a true-length view of the line of intersection AC. This is done by setting up the auxiliary plane 1 parallel to $a_F c_F$. The projection $a_1 c_1$ will be a true length.

Next select any point o_1 on the true length and pass a plane through this point perpendicular to the true length. This will be an edge view of a perpendicular cutting plane. Let this plane cut the edges of the original planes at x_1 and y_1. The plane $o_1 x_1 y_1$ is now a plane cutting the two planes at right angles to their line of intersection. Project back to the frontal view to locate the points o_F, x_F, and y_F.

The next step is to revolve the plane OXY in order to obtain its true shape. The plane OXY must be rotated until it is parallel to the frontal plane. We may choose any axis perpendicular to the frontal plane. In this case choose an axis through the point o_1. When the plane is rotated the points x_1 and y_1 will rotate in a circular path about o_1 to the positions x_{r_1} and y_{r_1} as shown. In the frontal view the points x_F and y_F will move perpendicularly to the axis to the positions x_{r_F} and y_{r_F}.

The true shape of the perpendicular cutting plane is shown as $o_F x_{r_F} y_{r_F}$. The dihedral angle is measured on this view as the angle $x_{r_F} o_F y_{r_F}$.

The horizontal projection has not been shown in Figure 11.19 for purposes of illustration. Note also that the analglyph Figure 11.19 has been turned around in order to provide a better view of the rotation of the cutting plane.

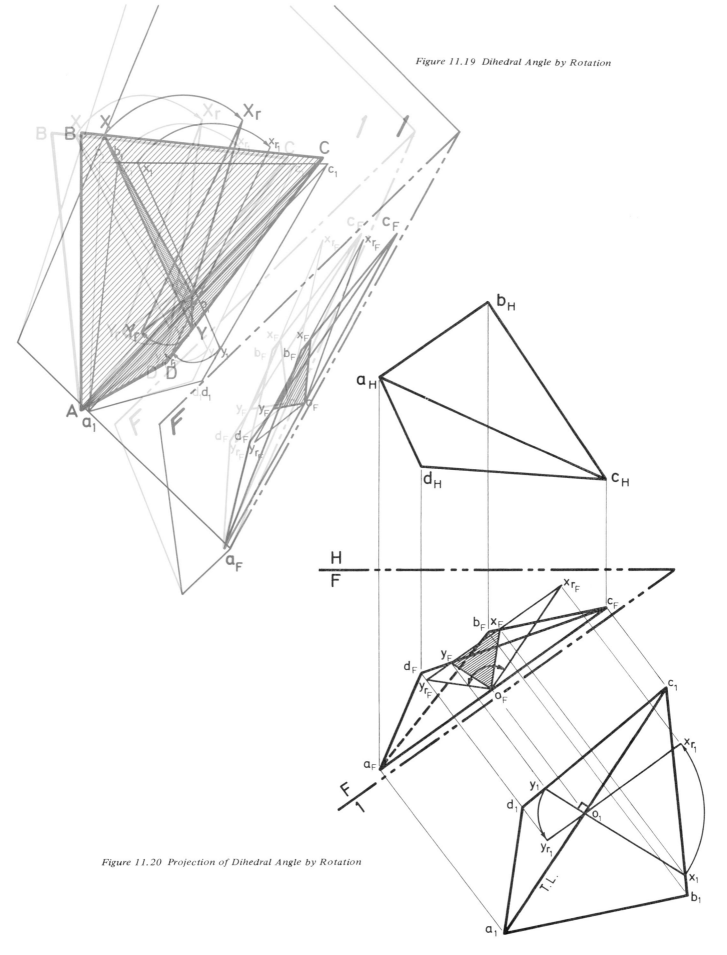

Figure 11.19 Dihedral Angle by Rotation

Figure 11.20 Projection of Dihedral Angle by Rotation

124

PROBLEMS

All dimensions in millimetres unless otherwise stated.

P 11.1 Find the true length of line AB by rotation. Also determine the angle it makes with the horizontal plane.

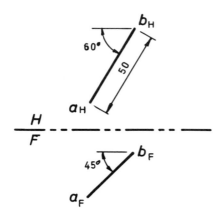

Figure P11.1

P 11.2 Revolve the line CD about the vertical axis MN until CD becomes parallel to the frontal plane. Measure the true length of CD.

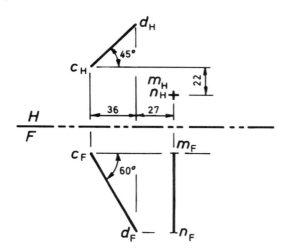

Figure P11.2

P 11.3 Find the true angle between the line AB and plane CDEF by rotation.

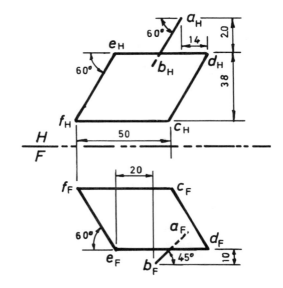

Figure P11.3

P 11.4 Draw a line 50 mm long making an angle of 30° with the horizontal plane and 45° with the frontal plane.

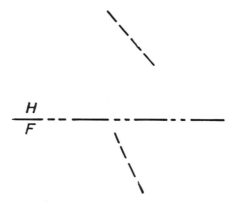

Figure P11.4

P 11.5 A 60° elbow and a 45° elbow joint are to be used to connect the vertical pipe to the horizontal pipe. Locate the centre line MN of the connecting pipe.

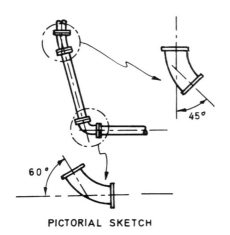
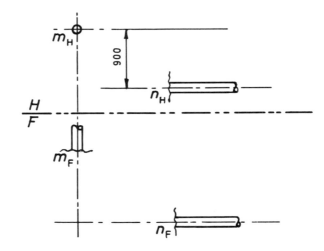

Figure P11.5

P 11.6 Obtain the true shape of plane ABC in the horizontal view by rotation.

P 11.7 Obtain the dihedral angle between faces A and B of the hopper. Choose a suitable scale.

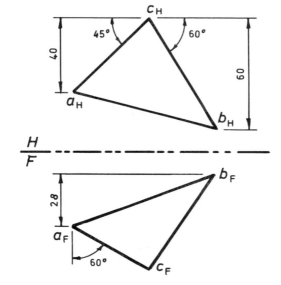

Figure P11.6

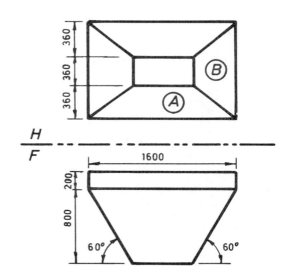

Figure P11.7

P 11.8 A cooling tower is to have the shape of a hyperboloid of revolution by revolving line AB about the line MN. Draw 24 equally spaced positions of line AB as it revolves. Show the H and F projections. What is the diameter of the innermost or gore circle of the hyperboloid?

P 11.9 Given the two views of the pipe elbow show in Figure P 11.9. Find the true angle between the planes containing the end flanges. Hint: choose a point P in space, drop perpendiculars to each plane from P (see Section 9.6). The angle between the two perpendiculars will be the required angle. Lay a piece of tracing paper over the drawing Figure P 11.9; trace enough of the two end flanges to locate them accurately in the H and F projections. There is no need to trace the whole elbow.

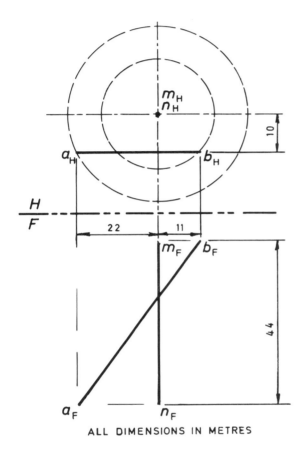

Figure P11.8

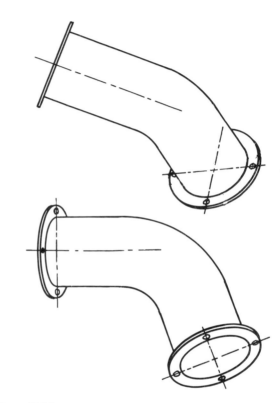

Figure P11.9

12. Pictorial Views

So far we have considered only orthographic projections in which the projection planes are perpendicular to each other and the point of sight is at infinity. In this situation the projection lines are parallel to each other and perpendicular to the projection plane.

It is sometimes desirable to produce drawings which are more pictorial in effect. This may be done by using projections which are not based on orthographic projection. Three such systems will be described, oblique projection, isometric projection, and perspective projection.

12.1 OBLIQUE PROJECTION

An oblique projection is obtained when the projection lines are *not* perpendicular to the projection plane. However the point of sight is assumed to be at infinity so that the projection lines are parallel to each other.

The comparison between orthographic and oblique projection is indicated in Figure 12.1. Note that the oblique projection lines all make the same angle θ_v with the projection plane which is a vertical plane.

The general situation, illustrating oblique projection is shown in Figure 12.2.

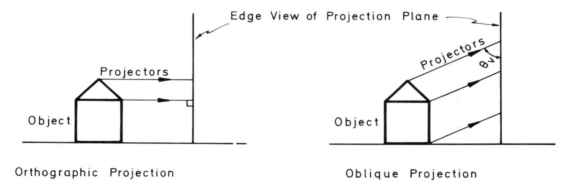

Figure 12.1 Orthographic and Oblique Projections

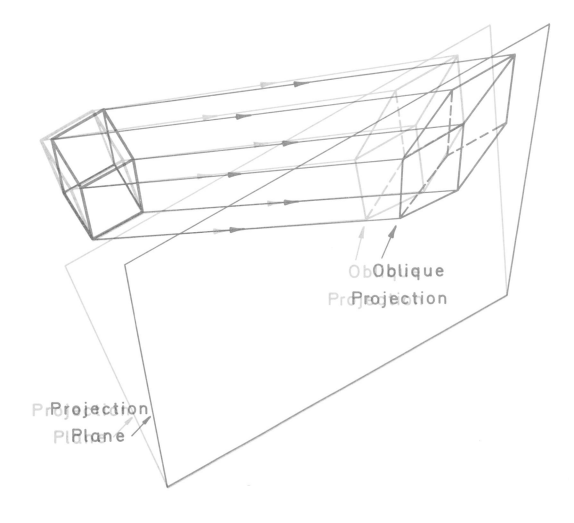

Figure 12.2 Oblique Projection

12.2 OBLIQUE PROJECTION FROM ORTHOGRAPHIC PROJECTIONS

The orthographic projections of an object are usually readily obtainable so it is convenient to be able to produce an oblique drawing of an object from its orthographic projections.

The first step is to decide on what you would like to see in the oblique projection. Do you wish a top view, a side view, or a view of the bottom? In addition there may be some distortion in the resulting view. For these reasons it will be shown that the appearance of an object depends on two conditions:

1. The angle θ_v of the projectors with the picture plane, and
2. The direction of the point of view at this angle.

Consider the situation shown in Figure 12.3, where the angle θ_v has been chosen to be 60°. Considering the point A of the object, a cone with a base angle of 60° will represent all the possible lines projected to the picture plane from A and at 60°.

A view along the line 1 for example will provide a top view and a left side view in addition to the view of the front of the object. Shown in Figure 12.3 are three other possible lines of projection 2, 3, and 4 each one of which produces a different view as shown in Figure 12.4.

Note that the front face which was placed parallel to the picture plane, is shown in its true shape, hence all dimensions are true lengths on the front face. Therefore, if an object has most of the important details on one particular face, then this face should be placed parallel to the picture plane in order to show it in its true shape.

The depth dimensions, however, will be foreshortened depending on the projection angle. The amount of foreshortening is explained in the next section.

Example 12.1 *Oblique Drawing Examples*

In the drawings below the face with the important or irregular detail is parallel to the projection plane.

Bell Crank

Bracket

Half Section Fitting

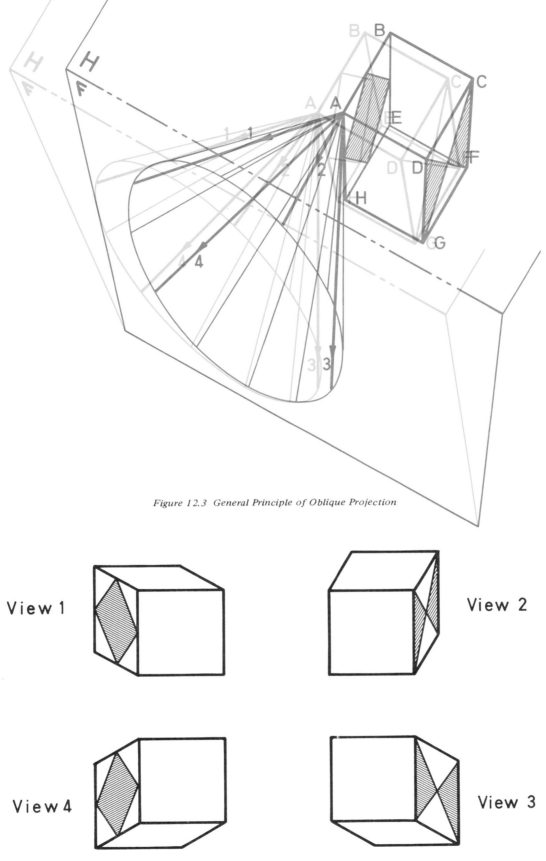

Figure 12.3 General Principle of Oblique Projection

Figure 12.4 Resulting Views

12.3 OBLIQUE PROJECTION PROCEDURE

Given the horizontal and frontal projections of an object as shown in Figure 12.5 we will construct an oblique projection.

First we must decide what projection angle to use. Suppose we select 60° and also we wish to view the object from above and to see the right-hand side DCGH.

From a convenient point on the horizontal projection, say point a_H, draw a triangle $a_H m_H n_H$ representing the horizontal projection of the cone. Angle $a_H m_H n_H = 60°$.

Draw the frontal projection of the cone which will be a circle of diameter $m_H n_H$ with centre at a_F. In this case select any projection direction above and to the right of a_F, and draw the line $a_F r_F$. Determine the horizontal projection of R, that is, r_H. Lines from the horizontal projection of the object parallel to $a_H r_H$ will determine the horizontal projections of all points on the object. Lines parallel to the direction $a_F r_F$ will determine the frontal projections of all points on the object.

It is now only necessary to find the piercing points in the frontal plane of the projecting lines determined above. The piercing point of the projection line from A is determined at the point a_N. Similarly other significant points on the object are found and joined to produce the required oblique view $a_N b_N c_N d_N e_N f_N g_N h_N$.

Foreshortening

The amount of foreshortening of the depth dimension may be determined as shown in Figure 12.6. The length of the projection of any line parallel to the projection plane is a true length. The length of lines perpendicular to the projection plane, such as the depth dimensions, is foreshortened according to the following relation. Foreshortened length KL = (true length KL) x cot θ_V.

Cavalier Projection

When the projection angle $\theta_V = 45°$, there is no foreshortening since cot 45° = 1.00. The type of drawing produced is called *cavalier* projection.

Cabinet Projection

In this case, $\theta_V = 63°26'$ and the foreshortening is 1/2 since cot 63°26' = 0.50. This is called a *cabinet* projection and sometimes produces a more realistic drawing than the cavalier type.

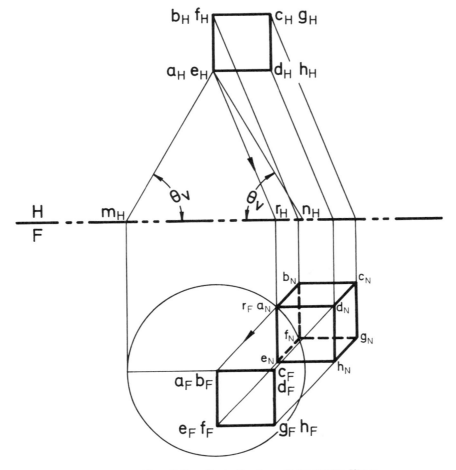

Figure 12.5 Oblique Projection from Orthographic Views

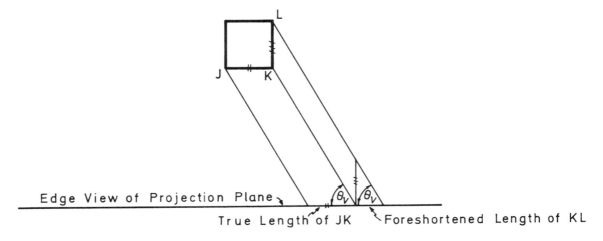

Figure 12.6 Foreshortening in Oblique Projection

12.4 ISOMETRIC PROJECTION

For isometric projection a projection plane is chosen so that it makes equal angles with all three principal planes. The point of sight is assumed to be at infinity so that the projectors are parallel. The general situation is shown in Figure 12.7 and 12.8.

In both illustrations, Figures 12.7 and 12.8, a plane ABC makes equal angles with the three principal planes. The isometric projection is formed on this plane.

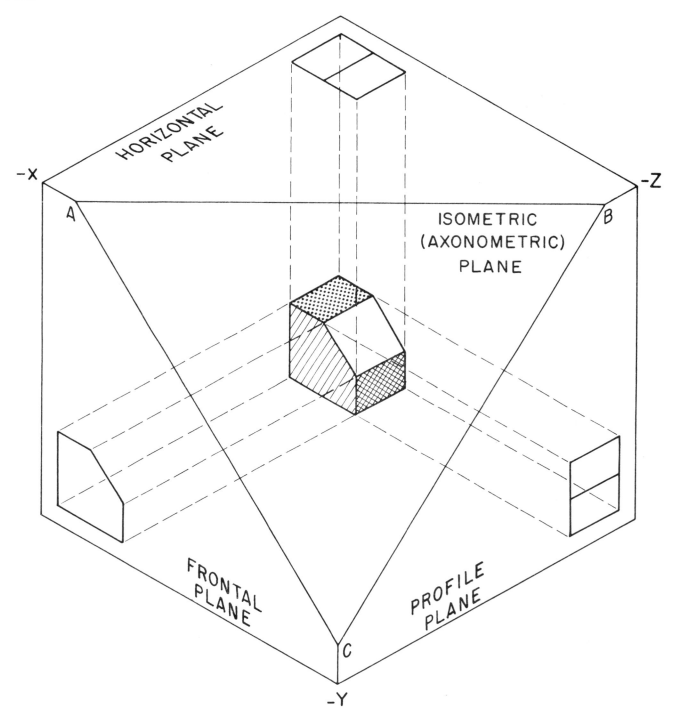

Figure 12.7 Isometric Projection Plane

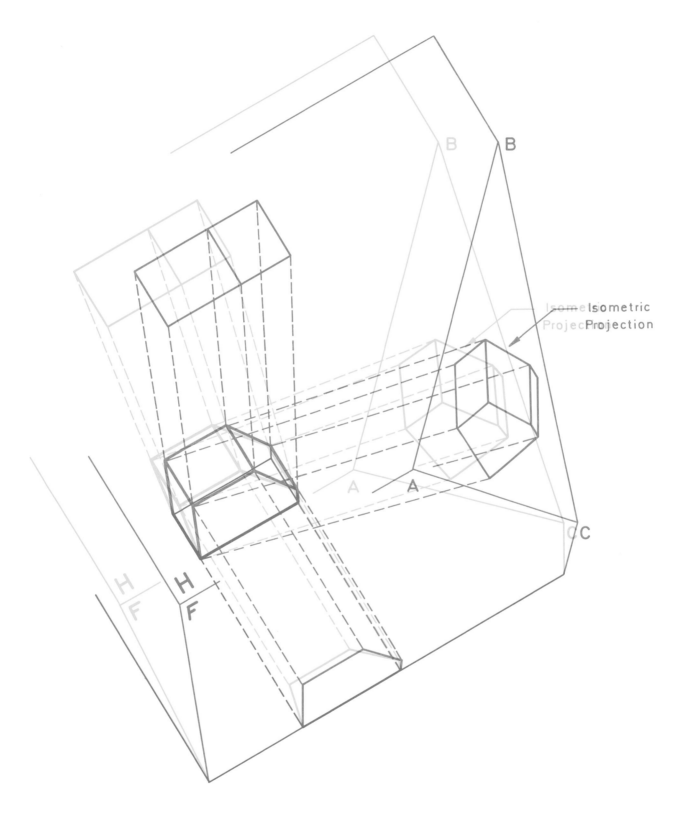

Figure 12.8 Isometric Projection Plane

12.5 ISOMETRIC VIEW FROM ORTHOGRAPHIC VIEWS

An isometric projection may be obtained from the standard orthographic horizontal and frontal projections.

The body diagonal of a cube will make equal angles with the faces of the cube. Therefore if we obtain a view looking along this diagonal it will be the required isometric projection.

Consider the block shown in Figure 12.9 in which we are given the H and F projections. A cube may be circumscribed about this block as shown by the broken lines. The body diagonal is the line DF. In order to look along this diagonal we require a point view of the true length. The procedure therefore is to obtain a true length of DF in plane 1 and then a point view of DF in plane 2. The projection plane 2 will provide the isometric projection of the cube and the block. The complete construction is shown in Figure 12.9. Note that an equal amount of foreshortening occurs for all lines parallel to the principal planes in isometric projection. The actual amount of foreshortening is

$$\frac{\cos 45°}{\cos 30°} = \frac{\sqrt{2}}{\sqrt{3}} = 0.816$$

The above procedure is somewhat lengthy but it illustrates the basic foundation of isometric projection. For most objects, however, a much simpler method will be explained in the next section.

Example 12.2 *Isometric Drawing Examples*

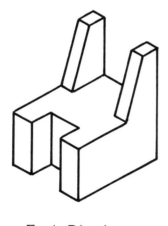

End Block

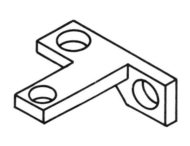

Support Guide

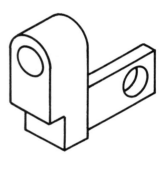

End Stop

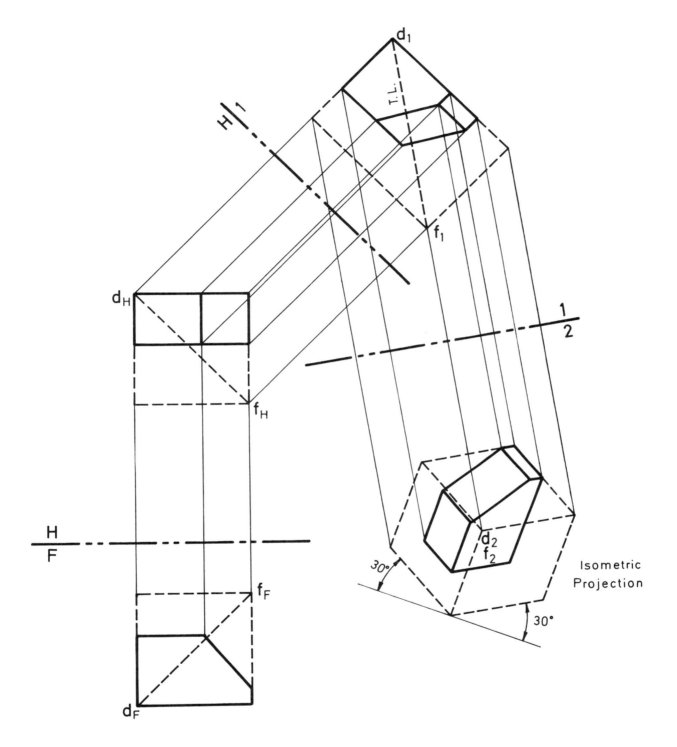

Figure 12.9 Isometric Projection from Orthographic Views

12.6(a) ISOMETRIC DRAWING—Normal View

The horizontal axes of a cube in isometric projection make equal angles of 30° with the horizontal as indicated in Figure 12.10. A simple method of making an isometric projection of an object is to set up axes at a convenient lower corner of the object and then scale off the position of significant points on these axes. A foreshortening of 0.816 on each dimension parallel to the axes should be used for a true isometric projection. However, the factor 0.816 merely changes the size of the drawing. Therefore it may be omitted entirely. The result will be an isometric *drawing* in contrast to an isometric *projection*.

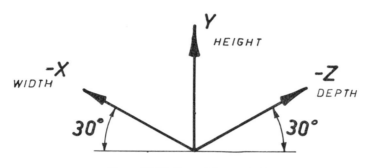

Figure 12.10 Normal Isometric Axes

The method is illustrated in Figure 12.11 where we are given the H and F projections of a cube and we wish to produce an isometric drawing. Set up coordinate axes on the H and F projections as shown. Draw a set of isometric axes at 30° as shown on the right of Fig. 12.11. Scale off the corner points of the cube along the isometric axes and join these points. The result is a normal isometric drawing. Note that the cube is somewhat larger than the orthographic views as we did not use the foreshortening ratio of 0.816.

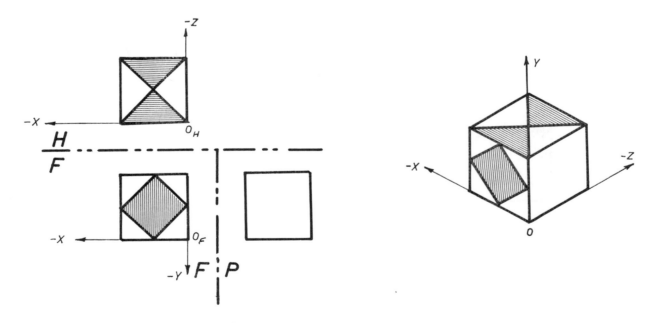

Figure 12.11 Normal Isometric Drawing

12.6(b) ISOMETRIC DRAWING—Inverted View

For some objects it may be more suitable to have a view from below. In this case inverted isometric axes can be used. In this situation the isometric axes are arranged as shown in Figure 12.12.

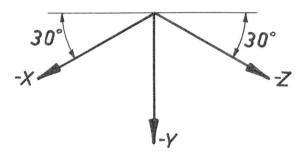

Figure 12.12 Inverted Isometric Axes

The method is illustrated in Figure 12.13 in which the axes are chosen to originate at a convenient top corner of the object. The height, y, dimension in this case is measured downwards. The resulting isometric drawing provides a view of the cube from below.

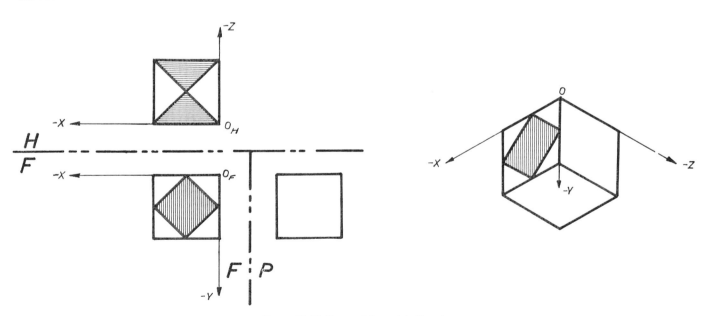

Figure 12.13 Inverted Isometric Drawing

12.7 CIRCLES IN ISOMETRIC

In isometric projections or drawings a circle which lies on one of the principal planes will appear as an ellipse. This is illustrated in Figure 12.14. Note that the minor axis is always at right angles to the major axis.

When the major and the minor axes are known an accurate ellipse may be drawn by the concentric circle method. In this method two concentric circles are drawn with diameters equal to the major and minor axes diameter as shown in Figure 12.15(a). To locate a point on the ellipse draw a series of radial lines one of which is shown as AB. This line cuts the circles at C and B. The point of intersection of a line through C parallel to the major ellipse axis will meet a line through B parallel to the minor axis at a point on the ellipse, D. Repeat the procedure to find a series of points which may be joined by a smooth curve to form the ellipse.

An approximate method of drawing an ellipse may be accomplished using a compass. In this method the centres for the small-end arcs are found at the intersection of the perpendiculars from the corners E and F of the circumscribing parallelogram as shown in Figure 12.15(b). The small arcs are drawn with radius r. The large arcs are drawn from the points E and F with radius R. For comparison an accurate isometric ellipse is shown dotted.

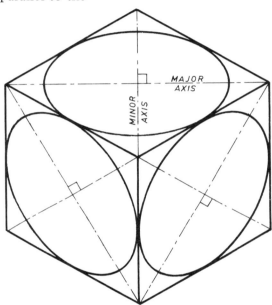

Figure 12.14 Isometric Ellipses

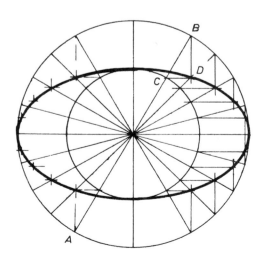

Figure 12.15(a) Accurate Construction

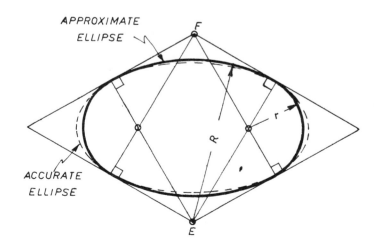

Figure 12.15(b) Approximate Construction

12.8 DIMENSIONING PICTORIAL DRAWINGS

Dimensioning of an oblique or isometric drawing often requires more care in placing the dimensions than for orthographic drawings. This is because there is only one view and thus the dimensions tend to become crowded.

There are two methods in general use. These are pictorial plane dimensioning and unidirectional dimensioning. In pictorial plane dimensioning the reference lines, dimension lines, and lettering all lie in one of the three pictorial planes. This is illustrated in Figure 12.16(a). In unidirectional dimensioning the lettering is all vertical and is made to read from the bottom of the drawing sheet. This is shown in Figure 12.16(b).

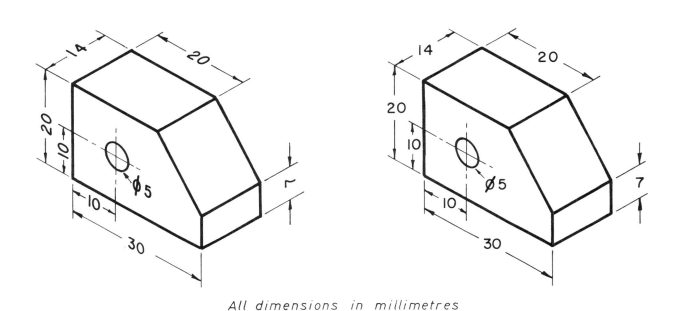

All dimensions in millimetres

Figure 12.16 Dimensioning of Pictorial Drawings

12.9 PERSPECTIVE PROJECTION

In perspective projection the point of sight is not at infinity as in oblique projection. The point of sight is placed at a finite distance from the object so that the projection lines to the picture plane are not parallel. This type of drawing can result in realistic views of an object as it simulates the views we see in daily life or in a photograph. The general situation is shown in Figure 12.17.

Some terms used in perspective projection are:

Station Point, S: the point of sight,

Picture Plane: a vertical projection plane on which the perspective drawing is obtained,

Horizon Line: the intersection of the picture plane with a horizontal plane through the point of sight, i.e., at eye level,

Ground Plane: a horizontal plane upon which the object is assumed to rest,

Ground Line: the intersection of the ground plane with the picture plane.

It can be seen from Figure 12.17 that the perspective projection of any point on the object is the frontal piercing point of the line from that point to the station point.

We will consider two methods of producing perspective drawings. The first method will describe the production of the perspective from the standard orthographic views of the object. The second method will make use of vanishing points.

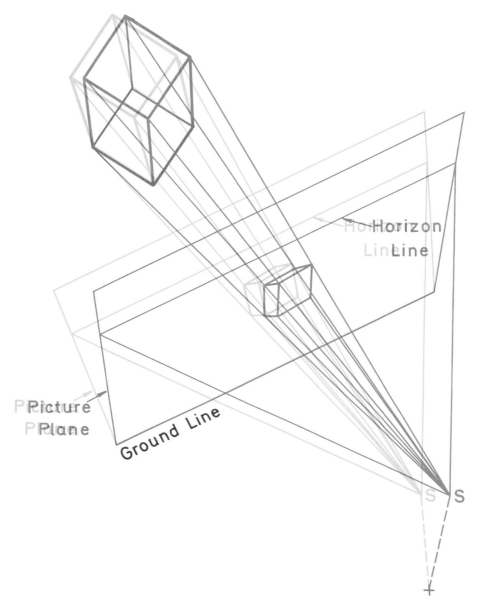

Figure 12.17 Perspective Projection

12.10 PERSPECTIVE PROJECTION FROM ORTHOGRAPHIC VIEWS

In this method the perspective projection is obtained from the frontal piercing points of lines or rays from a point on the object to the sight point S.

The situation is illustrated in Figure 12.18 in which the horizontal and profile plane projections only are shown as they are the only orthographic projections required. The horizontal and frontal projections could be used but the resulting perspective overlaps the frontal projection as these two projections are in the same plane.

The distance D of the sight point in front of the frontal plane and the height of the sight point are chosen to produce the required view of the object. For example you might wish to see the object from a high or a low position and to the right or left of the middle of the object.

A person with normal sight has a clear field of vision around the principal visual ray in the form of a cone of which the extreme angle is usually between 20° and 50°. However, the most distinct area is within an angle of 25° to 30°. For this reason the perspective drawing should be constructed within the angle of clear vision, unless an exaggerated view is required. In general a distance D between two and three times the diagonal of the perspective view should be used in order to avoid noticeable distortion. An even greater distance should be used to avoid any distorted appearance in the perspective.

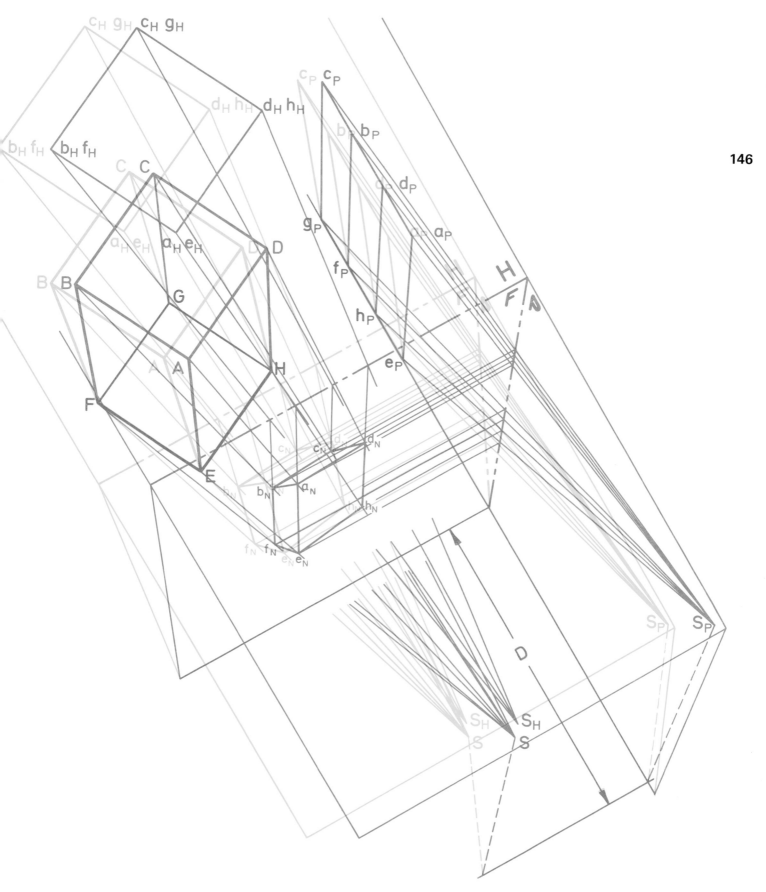

Figure 12.18 Perspective from Orthographic Views

12.11 PERSPECTIVE FROM ORTHOGRAPHIC
—Procedure

Given Data: The horizontal and profile plane projections of a cube and the station point.

Required: The perspective drawing of the cube.

Procedure: Note that both the horizontal and the profile projection of the station points S_H and S_P appear to be in the frontal plane but this overlap is due to the extension of the horizontal and the profile planes into this area of the paper. See Figure 12.19.

Draw the lines of sight in the profile plane from S_P to the corners of the cube a_P, b_P, c_P, d_P, e_P, f_P, g_P, and h_P. Note that this will determine the heights of the piercing points of these rays at the points where they intersect the right-hand edge view of the frontal or picture plane, i.e., the F/P line.

Draw the lines of sight in the horizontal plane from S_H to the corners of the cube a_H, b_H, c_H, etc. These lines of sight will intersect the top edge view of the picture plane, i.e., the H/ line, to locate the lateral position of the piercing points.

Project horizontally to the left from the height locations and vertically down from the lateral positions to locate the required perspective points a_N, b_N, c_N, etc. Connect the projected points with straight lines to construct the perspective view of the cube.

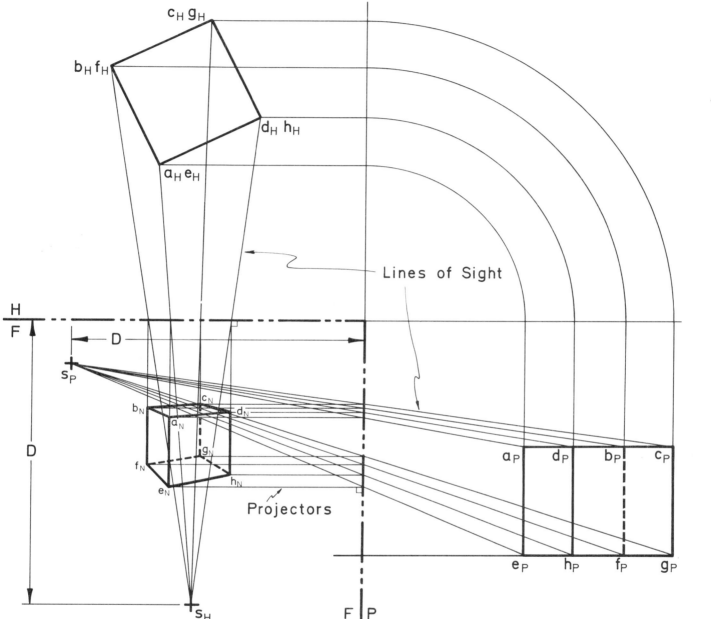

Figure 12.19 Perspective from Orthographic Views—Details of Procedure

12.12 VANISHING POINTS

The concept of vanishing points is based on the apparent convergence of parallel lines to infinity. In perspective projection the vanishing point is the point at which a horizontal line drawn through the station point parallel to one of the principal edges of the object intersects the horizon.

The situation is illustrated in Figure 12.20. In this case the vertical edges of the object are parallel to the picture plane and there are two vanishing points located by drawing lines of sight from the station point parallel to the two principal horizontal edges of the object. The intersection of these lines with the picture plane determines the location of the two vanishing points. The resulting perspective drawing is known as two-point perspective.

Note that parallel horizontal lines in the perspective projection converge to the vanishing points. Lines parallel to AD converge to the left vanishing point (VPL) and lines parallel to AB converge to the right vanishing point (VPR). This convergence provides a more convenient method of producing a perspective drawing. The procedure for determining the location of vanishing points is described in the next section.

Example 12.3 One-, Two-, and Three-Point Perspective of an Object

One-point perspective occurs when the major face of the object is vertical and parallel to the picture plane. Two-point has the face vertical but inclined to the picture plane. In three-point perspective all faces are inclined to the picture plane. These terms are sometimes called parallel, angular, and oblique perspective instead of one-, two-, or three-point respectively.

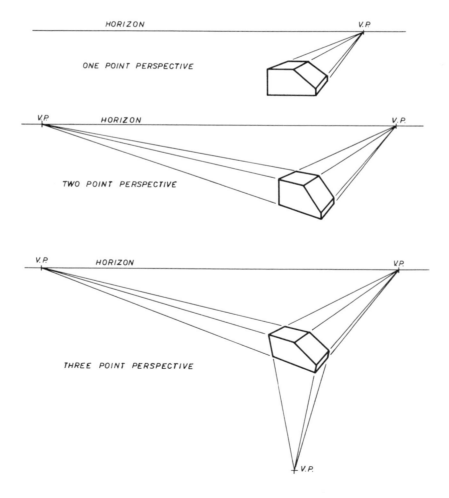

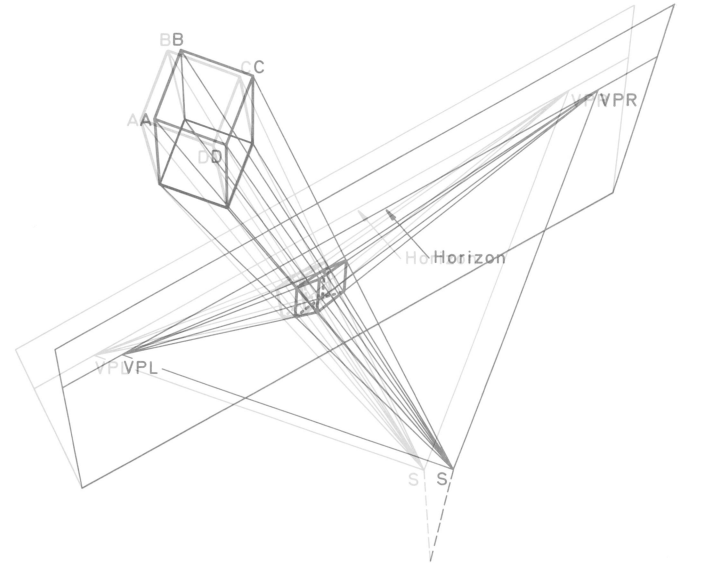

Figure 12.20 Vanishing Points

12.13 PROCEDURE FOR LOCATING VANISHING POINTS

Given Data: The horizontal projection of a cube and the station point.

Required: The left and right vanishing points for the principal horizontal lines of the cube.

Procedure: We wish to find the point where lines through the station point and parallel to the principal edges of the cube will pierce the picture plane. Therefore draw lines through S_H parallel to $a_H d_H$ and $a_H b_H$ to meet the top of the picture plane, as shown in Figure 12.21.

Draw in the horizon line on the picture plane at the height determined by the station point.

Drop a perpendicular from the point of intersection of the parallel lines with the top of the picture plane to intersect the horizon line at the vanishing points VPR and VPL.

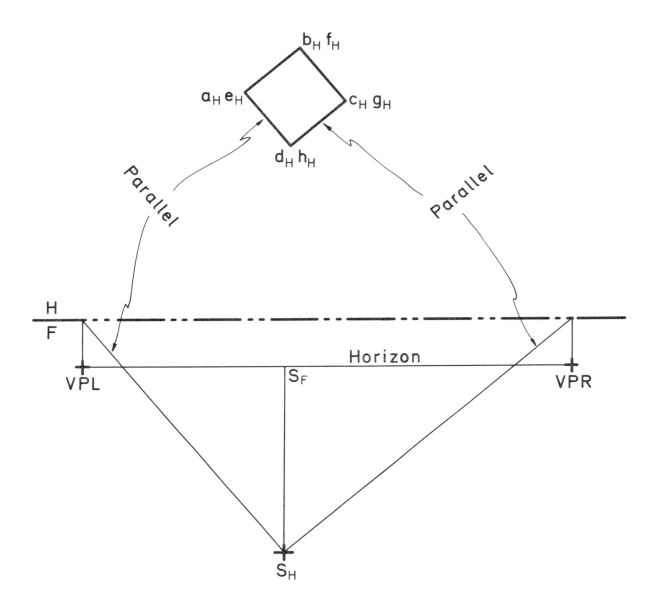

Figure 12.21 Locating Vanishing Points

12.14 PERSPECTIVE PROJECTION—Use of Vanishing Points

A perspective projection can usually be produced more quickly by making use of the vanishing points. However, in some cases they may lie at such a distance off the drawing board so that they are inaccessible.

The vanishing point was defined as the point where horizontal lines appear to meet at infinity on the horizon. In this method we make use of the fact that every horizontal line parallel to one of the principal edges of the object passes through the appropriate vanishing point for that edge. This establishes the perspective projection of the far-end point of these horizontal lines. The perspective of another point on the horizontal line can be found by determining the piercing point of the horizontal line with the picture plane. For example, the piercing point of BA is shown as PP_{ba}.

The general situation is shown in Figure 12.22. The procedure is detailed in the next section.

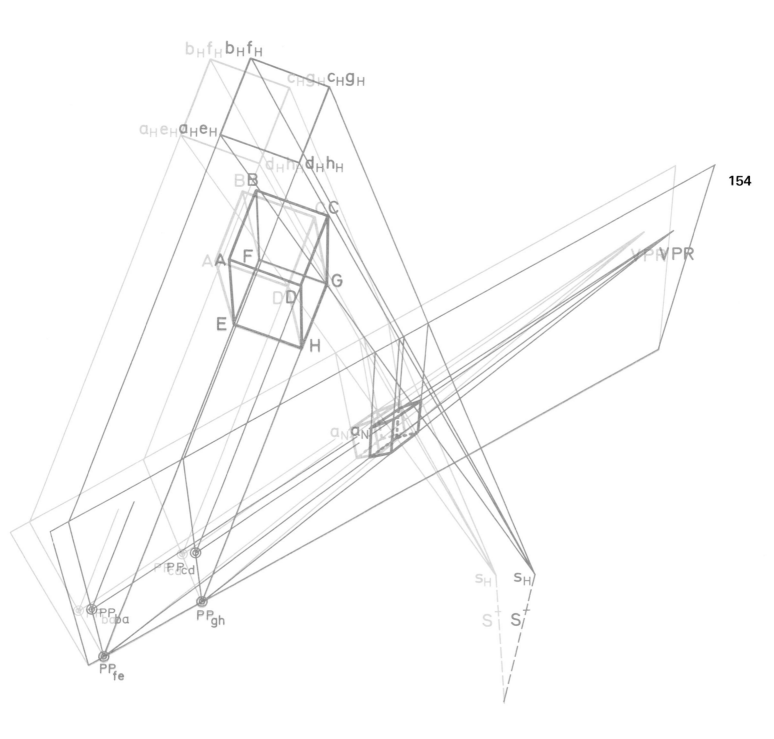

Figure 12.22 Use of Vanishing Points

12.15 PERSPECTIVE PROJECTION—Use of Vanishing Points—Procedure

Given Data: Horizontal projection of a cube and the height dimension. Location of the Station Point.

Required: Perspective projection of the cube.

Procedure: Locate the vanishing point of all horizontal lines parallel to AB at VPR, as shown in Figure 12.23.

Locate the piercing point of the horizontal lines on the cube which are parallel to AB, i.e., the lines BA, FE, CD, and GH. The piercing points are shown with double circles, PP_{ab}, PP_{cd} etc.

Join the piercing points with the VPR. The perspective projection of all horizontal lines parallel to BA lie on these lines.

Join the station point projection S_H to the horizontal projections of the corners of the cube, a_H, b_H, c_H, etc. The intersection of these lines with the top edge of the picture plane determines the lateral position of these points.

Draw perpendicular down from these intersections to meet the lines to VPR and thus establish the corners of the cube.

Join the projection of the corners a_N, b_N, c_N, etc. to produce the required perspective drawing.

It should be noted that alternatively we could have used VPL and the piercing points of lines BC, AD, FG, and EH to construct the required perspective projection.

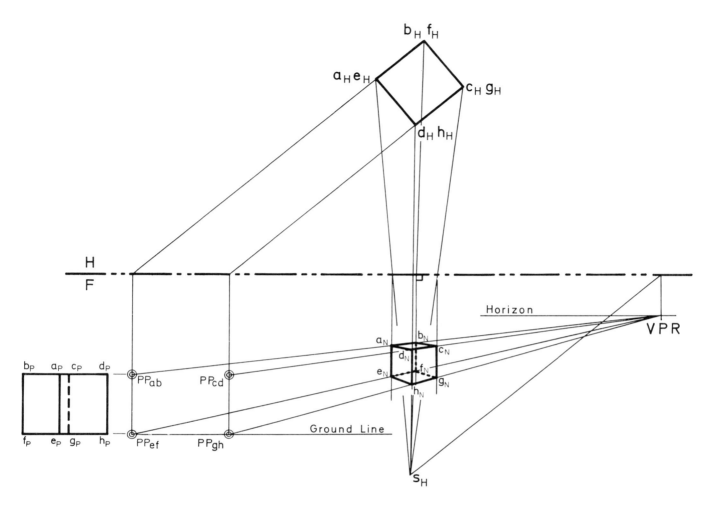

Figure 12.23 Use of Vanishing Points—Details of Procedure

12.16 PERSPECTIVE PROJECTION—General Procedure

It was noted previously that the vanishing-point method required the piercing points of horizontal lines. The location of the piercing points involved using the heights of points on the object.

A general procedure may be developed to produce a perspective projection by using the horizontal and either a frontal or profile projection of the object. Either of these latter projections will provide the necessary height data.

Given Data: The horizontal projection and the frontal projection of an object.

Required: A perspective projection of the object.

Procedure: Place the horizontal projection at a convenient distance near the top of the paper as indicated in Figure 12.24. Orient the projection to provide the view required.

Select a station point and locate its horizontal projection S_H.

Determine a vanishing point. In this case the left vanishing point was chosen VPL. Either the left or right could be used.

Determine the location of each significant point on the perspective projection. Join these points to produce the required perspective. An individual point will lie on the line joining its horizontal piercing point to VPL. The precise lateral location is found by drawing a perpendicular line from the intersection of the H/F line with the line from the particular point to the station point S_H.

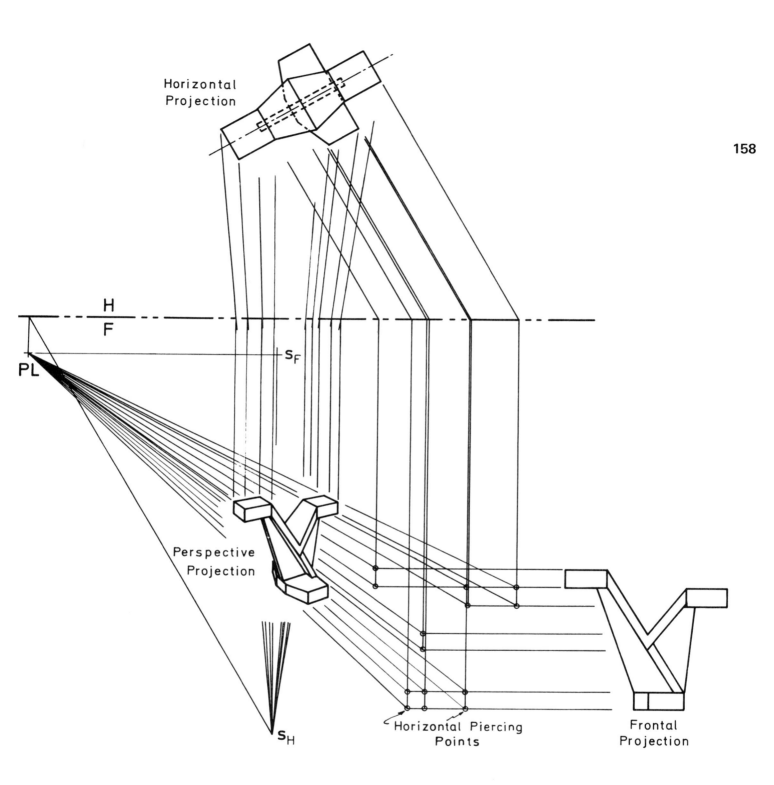

Figure 12.24 Example of Perspective Projection

PROBLEMS

Dimensions in millimetres unless otherwise specified.

P 12.1 Construct an oblique projection of the stepped block showing the top, front, and left side. Do the construction and final drawing from the orthographic views given in Fig. P 12.1.

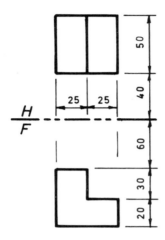

Figure P12.1

P 12.2 Construct an oblique cavalier projection of the ratchet wheel showing top, front, and right side.

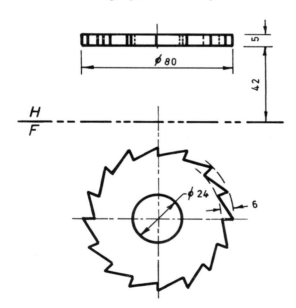

Figure P12.2

Oblique Drawing An oblique drawing may be made without going through the projection procedure. Start with a point representing the front corner of the object. Draw three axes, one vertical, one horizontal, and one at a convenient angle. On these axes plot the height, width, and depth of each point required and join to complete the drawing. This is similar to isometric drawing as explained in Section 12.6.

P 12.3 Make a cabinet drawing of the part shown.

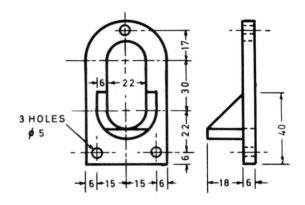

Figure P12.3

P 12.4 Make a cavalier drawing of the part showing the front, right, and bottom.

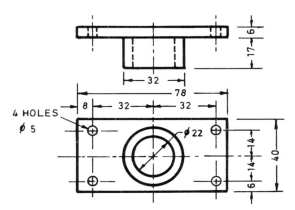

Figure P12.4

P 12.5 Produce an isometric drawing of the object shown.

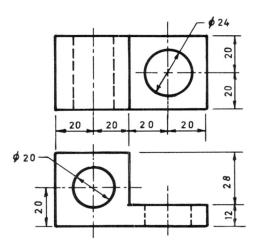

Figure P12.5

P 12.6 Make an isometric drawing of the object and dimension it completely.

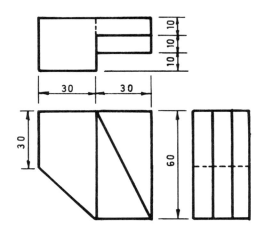

Figure P12.6

P 12.7 Make an isometric drawing of the object showing the bottom surface.

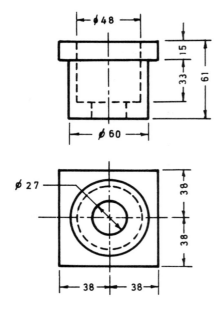

Figure P12.7

P 12.8 Make an isometric drawing of the concrete culvert. Locate a series of points on the curved surface by offsets from an enclosing rectangle. Plot the points along the isometric axes. Choose a suitable scale.

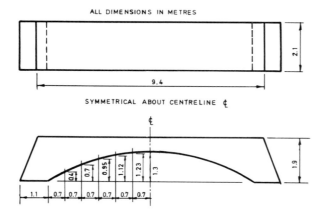

Figure P12.8

P 12.9 Construct a perspective view of the block shown in Figure P 12.9 and do not show hidden lines.

P 12.10 Construct a perspective view of the portion of the elevated highway of Figure P 12.10. Choose a suitable scale.

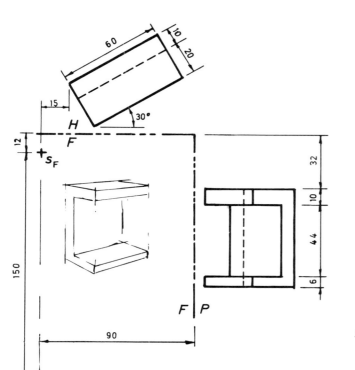

Figure P12.9

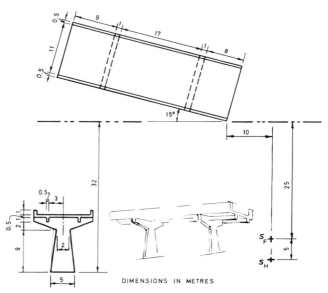

Figure P12.10

13. Vector Geometry – Coplanar Vectors

163 Many of the quantities used in engineering can be divided into two types, (1) scalar quantities which require only a magnitude to define them, and (2) vector quantities which require that both their magnitude and direction be specified. Examples of scalars are temperature and mass, and examples of vectors are force, velocity, and acceleration.

A vector quantity can be represented by a directed line segment, which has a definite length and an arrowhead at one end to indicate the sense of the direction or line of action. The length represents the magnitude of the quantity, to some scale, and the direction is along the line segment from the tail to the arrow head.

13.1 ADDITION OF VECTORS

An important requirement of vector quantities is that they must be combined or added in a certain manner which takes into account both the magnitude and direction. The method of addition of vectors is known as *the parallelogram law*. It can be shown experimentally that the result of adding two vectors together (say F_1 and F_2 in Figure 13.2) is equal in magnitude and direction to the diagonal of the parallelogram formed by the two vectors as sides. The sum of two or more vectors is called the *resultant*.

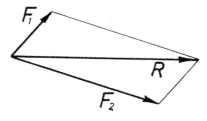

Figure 13.2 Parallelogram Addition

Figure 13.3 Triangle Addition

Figure 13.1 Graphical Representation of a Vector Quantity

In many cases, the addition of vectors can be more simply accomplished by using only half of the parallelogram as shown in Figure 13.3. This is known as *the triangle rule of addition.* The addition of several vectors can be done by repeated application of the triangle rule. For example, in Figure 13.4 the resultant of $F_1 + F_2$ is R_1. The vector R_1 is then combined with F_3 to form R_2 which is the final sum of the three vectors.

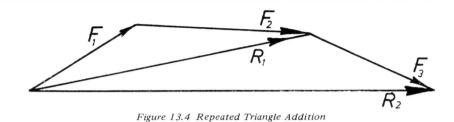

Figure 13.4 Repeated Triangle Addition

13.2 RESOLUTION OF VECTORS

Another useful concept in vector geometry is the idea that a vector can be resolved into any number of parts or components using the parallelogram or triangle rule. This is illustrated in Figure 13.5 where the vector V_1 is resolved into the two vectors V_2 and V_3. It can be seen that the vector V_1 can be replaced by vector V_2 and V_3 since the effect of V_2 plus V_3 is the same as the original V_1.

Figure 13.5 Resolution of Vectors into Components

13.3 COPLANAR VECTORS

In many problems the vectors, representing for example the forces or velocities, can be considered to lie in one plane, i.e., they are *coplanar.*

13.4 CONCURRENT COPLANAR VECTORS

In addition to being coplanar, some situations consist of a system of vectors which all meet at one point. Such systems of vectors are called concurrent.

13.5 SPACE DIAGRAMS AND VECTOR DIAGRAMS (concurrent coplanar vectors)

During the solution of problems it is convenient to use two types of diagrams, a "space" diagram showing the situation in regard to the physical aspects of the problem and the vector diagram associated with the problem. The space diagram is drawn to a space scale to show the physical configuration and accurate location of the points of application of the forces. The forces are not drawn to scale but their lines of action must be correct. The vector diagram shows the forces acting and of course must be drawn to a force scale. An example is shown in Figure 13.6: three weights connected by a cable are in equilibrium, and we are investigating the forces at point A.

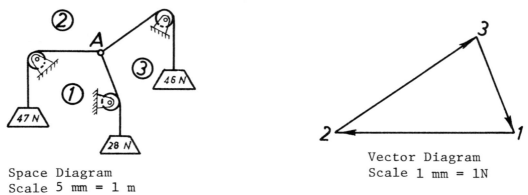

Space Diagram
Scale 5 mm = 1 m

Vector Diagram
Scale 1 mm = 1N

Figure 13.6 Space and Vector Diagrams

13.6 BOW'S NOTATION

In order to systematically keep track of the forces in vector geometry problems a certain notation as indicated in Figure 13.6, and called *Bow's notation*, will be used. In this method the spaces on both sides of every known or applied force are numbered consecutively in a clockwise direction around the system shown in the space diagram. These numbers are circled. In the vector diagram the vectors are drawn tip to tail starting at the vector between spaces ① and ②. This vector is identified as the vector 12 in the vector diagram. The first number denotes the tail end and the second number denotes the head or arrow end. The next vector 23 is added with its tail placed next to the tip of 12 as shown and then the final vector, 31, closes the triangle because this system is in equilibrium.

Another example is shown in Figure 13.7. In this problem the load is 4200 N and it is required to find the force in the strut AB and the cable BC. Since we are interested in the forces at B we number the spaces around the point B using Bow's notation.

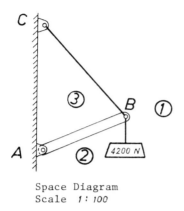
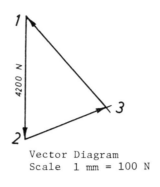

Figure 13.7 Another Example of Space and Vector Diagrams

The vector diagram is started by laying out the vector 12 parallel to the line of action between space ① and ② . This is drawn to scale. Next the vector 23 is drawn parallel to the line of action between ② and ③ . This vector is drawn to an undetermined length as its magnitude is not yet known, although the direction is known. Finally the vector 31 is drawn parallel to the cable and running from point 1 to meet the vector 23 at 3. At this time the vector diagram can be scaled to determine the magnitudes of all the forces. Note that the direction of vector 12 is pointing away from the point B indicating that this force is tension. Also note that the vector 23 is pointing towards the point B indicating that the force is pushing towards B and hence the member between ② and ③ is in compression. Similarly the direction 31 indicates tension in the cable. It is important to understand that the length of a member in the space diagram has no relation to the magnitude of the force it is transmitting.

13.7 EQUILIBRIUM

A rigid body is considered to be in equilibrium if it is at rest or moving with a constant velocity. Since the effect of a force is to change the linear velocity and the effect of a moment or couple is to change the angular velocity, then the body will be in equilibrium if the sum of all the forces acting on it are zero ($\Sigma F = 0$), and if the sum of all the moments acting on it are also zero ($\Sigma M = 0$). In the case of concurrent coplanar forces it is sufficient for equilibrium if the sum of the forces is zero since there is no moment arm to cause a product of a force and a distance, i.e., to produce a moment. In the more general force systems to be discussed both the above mentioned conditions for equilibrium must be satisfied, i.e., $\Sigma F = 0$ and $\Sigma M = 0$.

Consider the body shown in Figure 13.8 subjected to the action of the four forces as shown on the space diagram. These forces may be combined into a single resultant force by adding the forces vectorially as shown in the vector diagram.

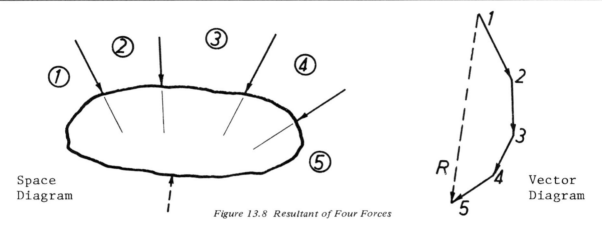

Figure 13.8 Resultant of Four Forces

The vector diagram provides the magnitude, direction, and sense of the resultant but it does not indicate the point of application of the resultant on the space diagram. The resultant is the force 15 in the vector diagram.

It should be noted that the body is not in equilibrium since the sum of forces acting is not zero (the vector diagram of only the four forces does not close). In order to satisfy $\Sigma F = 0$ a force 51 (opposite and equal to 15) must be applied to the body as shown by the dashed line on the space diagram. Note that the body will only be in equilibrium if the force 51 lies on the line of action of the resultant of the four original forces. This is the requirement to satisfy the condition that the sum of the moments must be equal to zero ($\Sigma M = 0$). If the force 51 on the space diagram has a different line of action from the resultant 15, a couple will result and the sum of the moments on the body will not be zero.

13.8 THE FUNICULAR POLYGON
In order to find the resultant of a system of forces and its line of action on the body, a method of resolving each force into two components is used in a graphic method called the funicular polygon in conjunction with the force polygon.

Consider a system of forces on a body as shown on the space diagram of Figure 13.9. It is required to determine the resultant of the forces and its line of action.

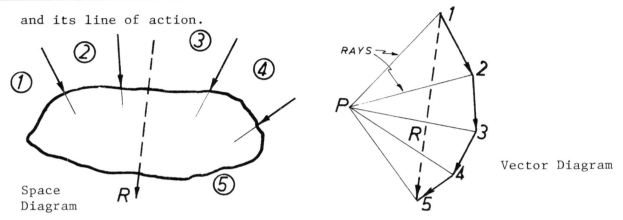

Figure 13.9 General Situation—Body with Four Forces

The force polygon is drawn as shown using Bow's notation on both diagrams. The resultant is the force 15. We have the magnitude and direction from the vector diagram (the force polygon) but we do not yet know its line of action on the space diagram. To find the line of action, each force is resolved into two components by drawing rays from any convenient point P to the ends of each vector. The vectors 1P, P2 are components of the force 12 and hence can be used to replace this force. This is illustrated in Figure 13.10 which shows two of the forces and their replacements.

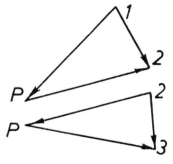

Force 12 is replaced by force 1P plus force P2. Similarly force 23 is replaced by 2P plus P3.

Figure 13.10 Replacement of Forces by Components

In a similar manner we can replace all the forces on the vector diagram by the forces to point P. These forces are called rays.

The next step is to replace the forces on the space diagram by these components, remembering that the forces on the space diagram show directions only and are not scaled as vectors. This is accomplished by starting at any convenient point on any force, say force 12 on the space diagram, and replacing the direction of force 12 by the directions 1P and P2 making use of the condition that the lines of action of the components of a force intersect on the line of action of the force they replace. This is illustrated in Figure 13.11 by the lines AB and AC parallel to 1P and P2 which replace the force between ① and ② . Similarly 2P plus P3 replace the next force 23. If we draw the components 2P and P3 through the point C we note that 2P is already drawn but acts in the opposite direction to P2. Therefore, we need to draw only the line CD. Similarly the line DE is drawn and finally the line EB parallel to P5 to intersect AB at B. The line of action of the resultant will pass through this intersection since it is the intersection of the lines of action of the two forces 1P and P5 which together equal the resultant R.

The lines AC, AB, etc., on the space diagram are called strings. The complete polygon ABEDCA is called the string, or funicular, polygon.

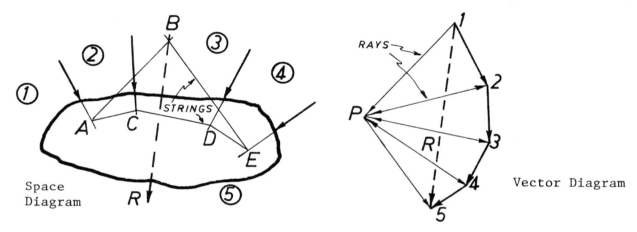

Figure 13.11 Resultant of Forces—Funicular Polygon

To review some of the procedures involved in finding the resultant and its line of action as detailed in the previous paragraph, consider the same problem with some of the lines omitted for clarity. The vector diagram of Figure 13.12 shows the resultant R as being the sum of the four forces in the vector diagram, i.e.,

12 + 23 + 34 + 45 = R.

Note that we can see from the vector diagram that the resultant is also the sum of two of the components, i.e.

1P + P5 = R.

All the intermediate pairs of forces P2, 2P; P3, 3P; and P4, 4P, cancel out because for each pair one is equal and opposite to the other. On the space diagram we can see that the resultant R is drawn through the intersection of the two lines of action AB and EB which are parallel to 1P and P5.

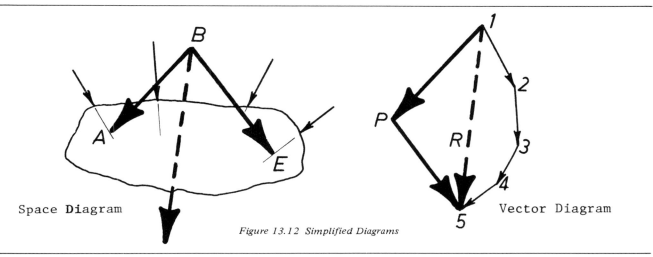

Figure 13.12 Simplified Diagrams

Space Diagram   Vector Diagram

The above diagram is to emphasize the statement made earlier that the lines of action of the components of a force intersect on the line of action of the force they replace. In this case the two components of the resultant R are 1P and P5 and their lines of action are AB and EB. The intersection of AB and EB at B determines the point through which resultant R must pass on the space diagram.

13.9 COPLANAR PARALLEL FORCES

The line of action of the resultant of a system of parallel forces is found by a straightforward application of the string polygon method. This is illustrated in Figure 12.13 in which the force polygon becomes a straight line. Note that in this case the pole point P is taken to the right of the vector polygon as it can be placed in any convenient position.

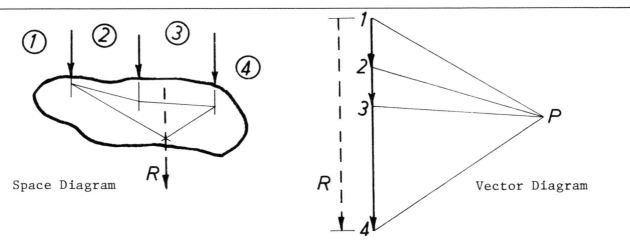

Figure 13.13 Resultant of a System of Parallel Forces

Example 13.1 Resultant Force

The electric power utility pole shown has four forces acting at the top. Find the resultant of these forces.

Solution The lengths of the vectors on the space diagram have no significance. They show direction only. Assume a direction for the resultant and sketch it on the space diagram. Identify the forces using Bow's notation, ①, ②, ③, ④, and ⑤.

Select an origin for the vector diagram at point shown +. From this origin draw vector 12 parallel to its direction in the space diagram. Scale this vector using 1 mm = 10 N and number it 12. Place an arrow at end 2 to show the direction.

Add vectors 23, 34, and 45. The resultant is 15. Draw this resultant force on the space diagram in a direction parallel to 15 on the vector diagram. Scale the resultant 15 on the vector diagram.

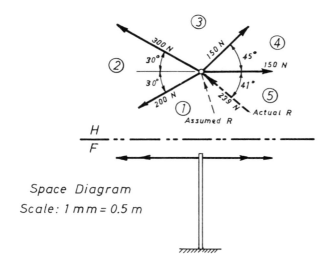

Space Diagram
Scale: 1 mm = 0.5 m

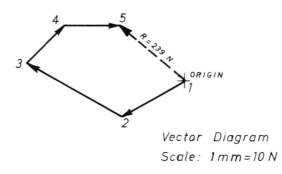

Vector Diagram
Scale: 1 mm = 10 N

13.10 RESULTANT COUPLE

In some situations, while the sum of the forces on a body is zero, the sum of the moments may not be, in which case the body is not in equilibrium. This condition can be determined from the funicular polygon method. For example the body shown in Figure 13.14 is under the action of a system of forces which form a closed force polygon, therefore $\Sigma F = 0$. Drawing the funicular or string polygon as shown, it is found that the two components 1P and P1, being components of the original forces 12 and 51 respectively are not colinear so that they form a couple with a moment arm "d". Therefore, the resultant is a couple of magnitude (1P) x d in a counterclockwise direction.

The magnitude of the couple is the product of a force and a distance. The force 1P is scaled from the vector diagram as so many newtons (or pounds) force. This distance "d" is scaled from the space diagram as so many metres (or feet) so the answer is in newton metres or foot pounds. The direction of the couple must also be stated, clockwise or counterclockwise. In this case the couple is counterclockwise.

This body could be in equilibrium under the action of the given forces if the line of action of one or more of the forces is shifted to a position so that the distance "d" becomes zero. For example the force 51 could be shifted to the left until it passes through the point E as shown. It can be seen that for a body to be in equilibrium both the force polygon must close and the string polygon must close.

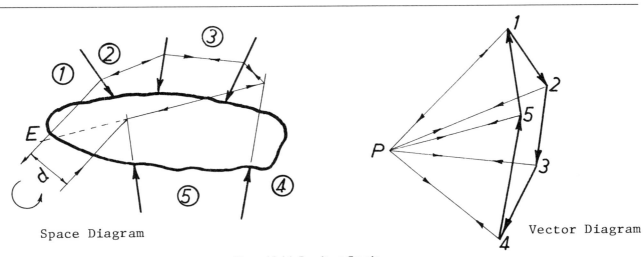

Space Diagram   Vector Diagram

Figure 13.14 Resultant Couple

13.11 BEAMS

A beam may be defined as a long member used to carry loads over an extended distance. Familiar examples are roof and floor beams, car axles, shafts in motors, and lathe and milling machine spindles. The general situation is that the beam is loaded at several points along its length and is supported at each end. Some examples are shown in Figure 13.15 where "F" indicates an applied load and "R" indicates the end or bearing support force. These support forces are usually called reaction forces as they are a result of (a reaction to) the applied forces.

In these systems the loads are usually known (to a certain degree of accuracy) and it is required to find the end reactions. It is apparent that the problem is similar to that of finding the resultant of forces using a funicular polygon as described previously. The main difference is that we now have two unknown forces, that is the two end reactions, instead of the one unknown resultant in the previous examples.

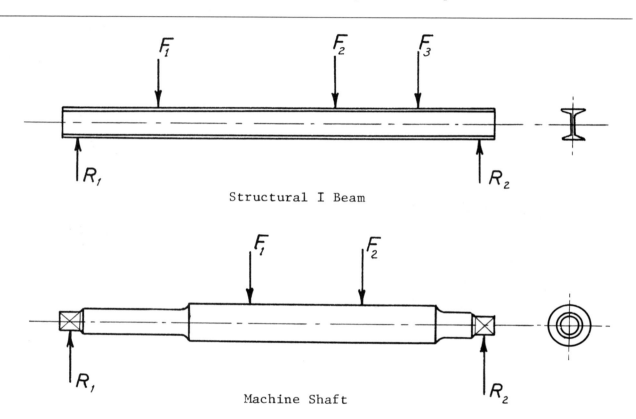

Structural I Beam

Machine Shaft

Figure 13.15 Examples of Beams

Consider the problem of determining the end forces required to support the gear shaft shown in Figure 13.16.

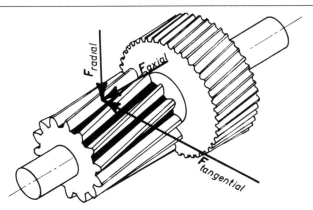

Figure 13.16 Forces on a Helical Gear

For simplicity consider the forces in the vertical plane only. The radial and axial forces for each gear may be combined so that the space diagram will be as shown in Figure 13.17. Since the loads are inclined a bearing will be required at one end of the shaft to take the horizontal thrust load in addition to the vertical radial load. Assume that the left-end bearing can take both thrust and radial loads and that the right-end bearing can take only radial loads. Therefore we know that the direction of the right-end reaction is vertical but we do not know the direction of the left-end reaction (at this stage). However we do know that the left-end reaction is applied at point A.

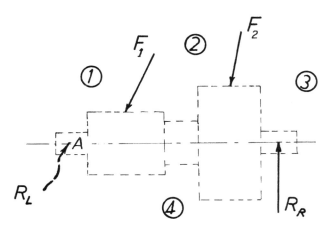

Space Diagram
Scale 1 mm = 5 mm

Figure 13.17 Reactions on a Gear Shaft

The solution will be explained in a step-by-step procedure.

Step 1 Draw the space diagram to a scale 1 mm = 5 mm. This diagram is a schematic of the physical problem. Use Bow's notation to identify all the forces.

Step 2 Draw the vector diagram to a force scale of 1 mm = 100 Newtons. Start with force F_1 and draw each force in turn going around the space diagram in a clockwise sequence. That is draw force 12 and then add force 23 to it on the vector diagram. The next force is the reaction R_R for which we know only the direction, which is vertical. Therefore draw a vertical line through point 3 on the vector diagram. This is as far as we can go in adding forces on the vector diagram at this stage.

Step 3 Select a convenient location for the pole point "P" and draw in the rays P1, P2, and P3 on the vector diagram. (You can't draw in ray P4 because you haven't got a point 4 yet).

Step 4 Go to the space diagram. Select a starting point for the string polygon at the point where the unknown left-end reaction is applied to the shaft, point A. Through this point draw the 1P component of force 12, that is a line parallel to 1P to intersect the line of action of force 12 at B. The other component of force 12, that is, ray P2 is drawn through B to intersect the next line of action at C. Similarly D is found on the line of action of R_R. Join D to A to close the polygon.

Step 5 Go back to the vector diagram. Draw ray P4 parallel to AD to intersect the vertical line through point 3 at point 4. Join 41. The vector diagram can now be scaled to obtain the values of R_R and R_L.

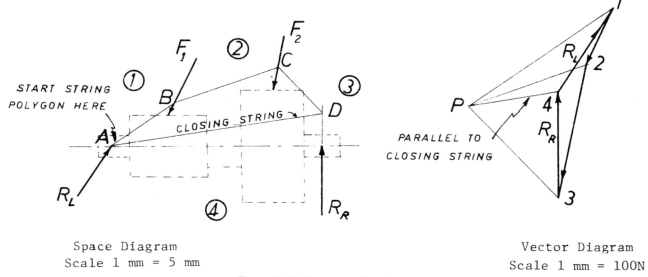

Figure 13.18 Forces on a Gear Shaft

13.12 TRUSS ANALYSIS

A structural truss is actually a very long, deep beam. It is commonly used in bridges, roofs, and special machine structures. Various standard types of truss are shown in Figure 13.19. The loads acting on a truss may be due to the structure itself, live loads such as vehicles, wind loads, and snow loads. Knowing the load and the type of end support, it is possible to analyse the truss to determine the force in each member. In addition we can ascertain whether each member is in compression or tension.

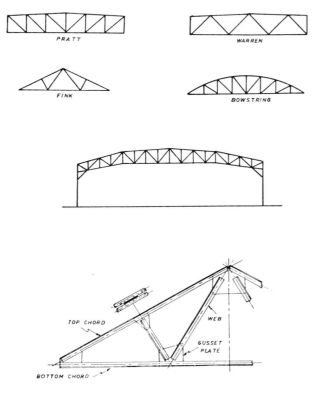

Figure 13.19 Structural Trusses

13.13 DETERMINE THE TRUSS END REACTIONS

The first step is to consider the truss to be a beam and determine the end reactions.

As an example, the truss shown in Figure 13.20 will be analysed. The individual members would likely be rivetted or bolted together, however we consider them to be pinned as shown by the small circles at each end.

Using Bow's notation, number the spaces between the loads as shown and construct the funicular polygon. The space between the line of action of R_L and the left end load is zero. Therefore the "string" P/1 of the polygon will be of zero length. That is, when we choose the starting point of the equilibrium polygon on the line of action of R_L, we have automatically drawn string P/1. Likewise P/6 will also be of zero length.

Example 13.2 *End Reactions*

The latch beam shown below is loaded by the three forces 200, 500, and 500 N. Find the reactions at A and B.

Solution Identify all the forces using Bow's notation ①, ②, ③, ④, ⑤.

End B is a roller type of support designed to allow movement due to temperature changes. The direction of the force at B is normal to the rolling surface.

The direction of the force at A is unknown. Therefore, the funicular polygon must be started at point A as it is known that the force 51 passes through point A. Proceed as described in Section 13.11.

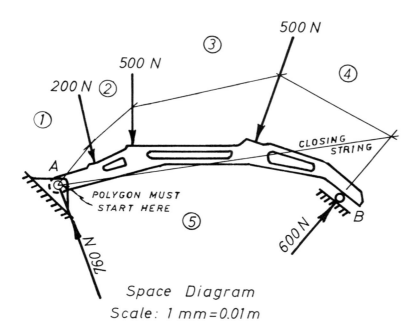

Space Diagram
Scale: 1 mm = 0.01 m

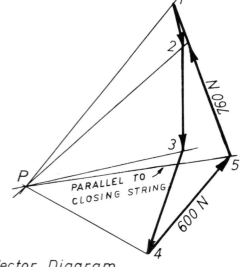

Vector Diagram
Scale: 1 mm = 20 N

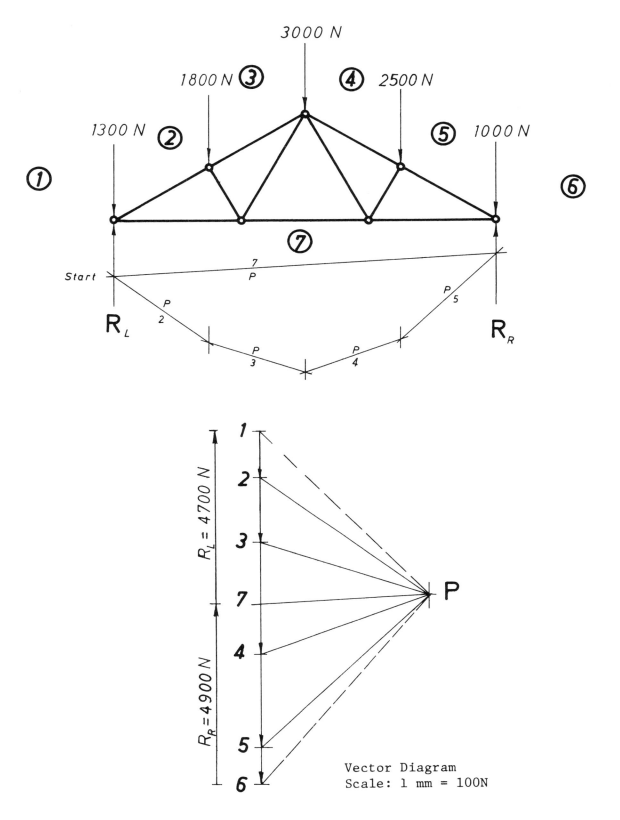

Figure 13.20 Fink Truss Example

13.14 THE METHOD OF JOINTS

The next step is to consider the forces in each member of the truss. To do this each member must be identified. Therefore label each panel of the truss with a lower case letter a, b, c, etc. as shown in Figure 13.21(a). In addition each panel point is lettered with an upper case letter A, B, C, etc.

Each joint or panel point will be considered by itself. This is known as the *method of joints*. When a joint is isolated it is considered to be a "free body" and therefore if it is in equilibrium then the sum of the forces and moments acting on it must be zero.

To solve for the forces at a joint it is necessary to start at a panel point which has not more than two unknown forces. In this example points A and G are the only ones fulfilling this condition. Consider the point A and the free body diagram as shown in Figure 13.21(b). The forces acting on this point are the reaction 71, the load 12, and the forces in the members 2a and a7. Since these forces are in equilibrium we can draw the force polygon as shown in Figure 13.21(b). To draw the polygon start at the convenient point ⑦ and draw the vector 71 parallel to R_L and to the same scale as the vector diagram of Figure 13.20. Put an arrow head on this vector as shown to indicate its direction. Add the vector 12 to the end of vector 71, obtain the point ②. To complete the polygon, which must close, add the vector 2a, parallel to the member AB, to the end of the vector 12. We do not know the length of this vector as yet. The closing vector will have to be the last force a7 drawn through the point ⑦ and parallel to the member CA. Add the direction arrow heads to these last vectors.

The vector diagram for the point A is now complete and from it we can determine whether the members AB and AC are in tension or compression. From the vector diagram the arrowhead on the force 2a is pointing from right to left. Transfer this arrow to the free body diagram of Figure 13.21(b). Its direction indicates that member AB is pressing towards the point A, therefore the member must be in compression. In a similar manner the member AC will be in tension since the force a7 is pulling away from A.

Now that the forces in the members AB and AC have been found it is possible to analyse the next two joints B and C. The free body diagram for the joint B and its vector diagram are shown in Figure 13.21(c). We can determine the direction of the force a2 by considering the member AB as a free body. For equilibrium the force a2 must be equal, opposite, and colinear with 2a. Start at the point a and draw the vectors a2 and 23. Close the polygon with the vectors 3b and ba.

In a similar manner the forces at each joint of the truss are found one joint at a time.

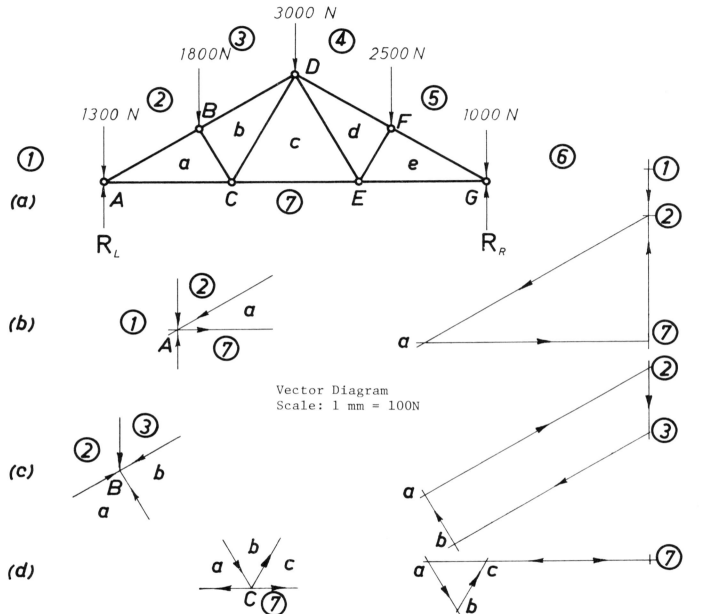

Figure 13.21 Fink Truss Example Continued

It can be seen that all these vector diagrams could be combined into a composite vector diagram placed on the original load line in Figure 13.20. This is shown in Figure 13.22. Since each vector would have two arrow heads running in opposite directions they are omitted. The "sense" they conveyed is given by the way we read the vector when we go around the point, under consideration, in a clockwise direction. The "free body" diagrams can be dispensed with since all the information they contain may be obtained directly from the truss drawing (situation diagram).

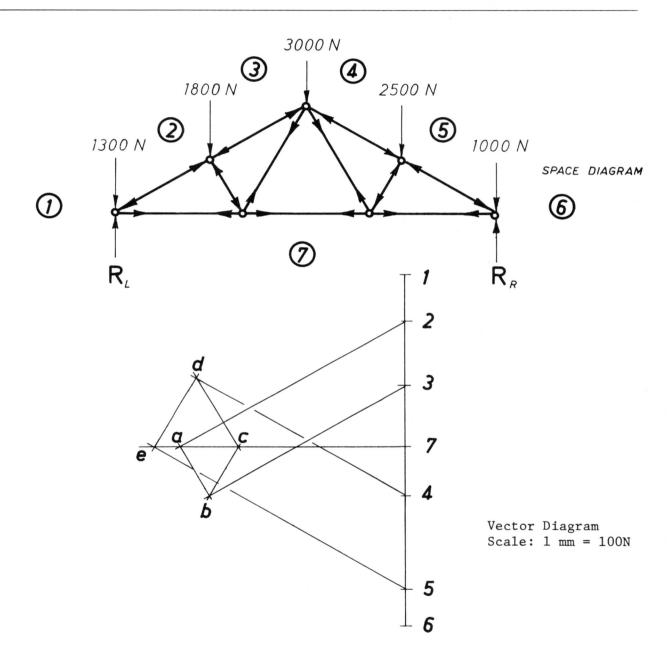

Figure 13.22 Fink Truss Example Continued

13.15 EQUILIBRIUM OF COPLANAR NONCONCURRENT FORCES

For any set of coplanar nonconcurrent forces to be in equilibrium the sum of the forces and the sum of the moments must be zero ($\Sigma F = 0$ and $\Sigma M = 0$). This means that if we divide the forces into two groups, the *resultant* of all the forces in one group must be equal in magnitude, opposite in direction and colinear with the *resultant* of the forces in the second group.

This fact is very useful in the case where a known force is to be held in equilibrium by three forces whose directions only are known. This situation is illustrated in Figure 13.23. Here a force F = 5N acting in the direction shown is to be held in equilibrium by the forces A, B, and C whose directions only are given.

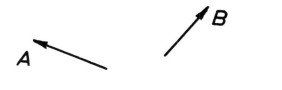

Figure 13.23 Four Forces

At this stage it is not possible to draw the vector diagram $\Sigma F = 0$ (i.e., $A + B + C + F = 0$) since more than two quantities are unknown. There are three unknowns, the magnitudes of the forces A, B, and C.

Using the principle of dividing the forces into two groups, consider one group consisting of A and B, and the other being C and F. The resultant of A and B must pass through the intersection S of their lines of action as shown in Figure 13.24. Similarly the resultant of the other group must pass through the point T.

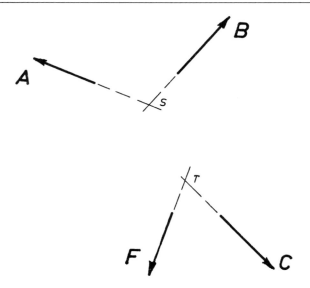

Figure 13.24 Four Forces Grouped

The two resultants, R_{AB} and R_{FC} must be equal, opposite, and colinear for equilibrium. Therefore the resultants must have the same line of action, along the direction ST. Figure 13.25(a).

The vector diagram can now be drawn as shown in Figure 13.25(b). Start by drawing the force F to some convenient scale. Add the force C to F. Draw in the resultant force R_{FC} parallel to ST to meet force C at W. Now add the forces A and B to cause $R_{AB} = -R_{FC}$ and close the complete polygon.

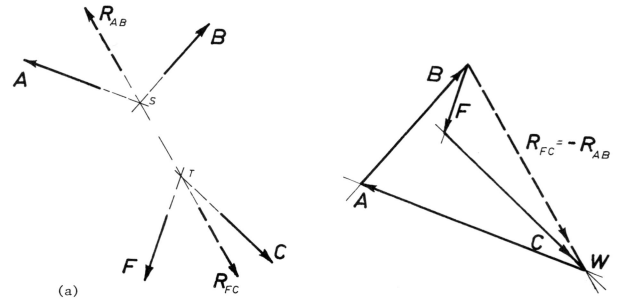

Figure 13.25 Solution—Four Forces

13.16 THE METHOD OF SECTIONS

The method of joints was used previously to find the forces in each of the members of a truss. This method may involve extra work if the forces in only a few of the members are required. The method of sections can be used to reduce the effort required. The following example will illustrate the method.

Self-supporting towers like the one shown in Figure 13.26 must be designed to withstand wind loads as well as to carry any microwave or telecommunication equipment at the top. These external forces cause the loads in the members to increase as you move from the top to the bottom of the tower. Thus the size of the members should increase continuously in order not to waste material. However, this leads to problems in construction. As a compromise, the tower will often be divided into sections. Within any section, all diagonals will be the same size and all vertical members will be the same size. For each section, these sizes are determined by the lowest members in that section because these are the ones carrying the greatest load.

For the tower shown in Figure 13.26(a) the designer really only has to know the forces in the members at levels m—m and n—n. To find them, it is necessary to consider two free bodies, one of the section above m—m, and one of the sections above n—n. The forces in the cut members must now appear as external forces. You can now see that in both Figure 13.26(b) and 13.26(c) there are three vectors whose directions are known but whose magnitudes are unknown. If we now find the

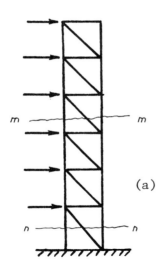
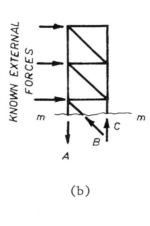
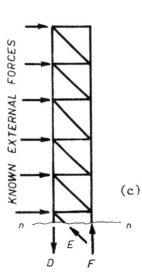

Figure 13.26 Self-Supporting Tower

resultant of the other external forces acting on each free body, we get the situation shown in Figures 13.27(a) and 13.27(b). If you compare these figures with Figure 13.25, you will see that the graphical method used to find the magnitude of the forces in that case works just as well here. The vector diagrams obtained are shown in Figures 13.27(c) and 13.27(d).

Remarks:
1. When you cut the body across the members whose forces you are trying to find, you can take a free body of either of the two sections. The one to take is the one that looks to be the more convenient. However, either one will give the same result.

2. For concurrent coplanar forces in equilibrium, $\Sigma M = 0$ automatically. Therefore, two unknowns are the most that can be determined. This corresponds to graphically solving two equations $\Sigma F_x = 0$, $\Sigma F_y = 0$.

3. For nonconcurrent coplanar forces in equilibrium, we can solve problems with three unknowns. This involves the graphical solution of the three conditions $\Sigma F_x = 0$, $\Sigma F_y = 0$, $\Sigma M = 0$.

Example 13.3 *Application of Method of Sections*
The steel casting with centre of gravity at G is supported at corners A, B, and C by rollers. Find the three corner reactions.

Solution Because of the roller supports the directions of the three reactions are known. The method of sections can be used.

Draw the lines of action of the forces. R_B and W intersect at G, and R_A and R_C will intersect at N. The resultant of each pair must lie along the line GN.

Start the vector diagram by drawing the weight W. Add R_B to complete the triangle of W, R_B, and resultant R.
Similarly combine R_A and R_C to form R.
Scale the vectors R_A, R_B, and R_C.
An alternate solution could be obtained using the intersections of R_B and R_C, and of W and R_A.

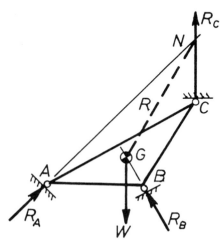

Space Diagram

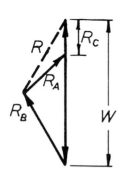

Vector Diagram

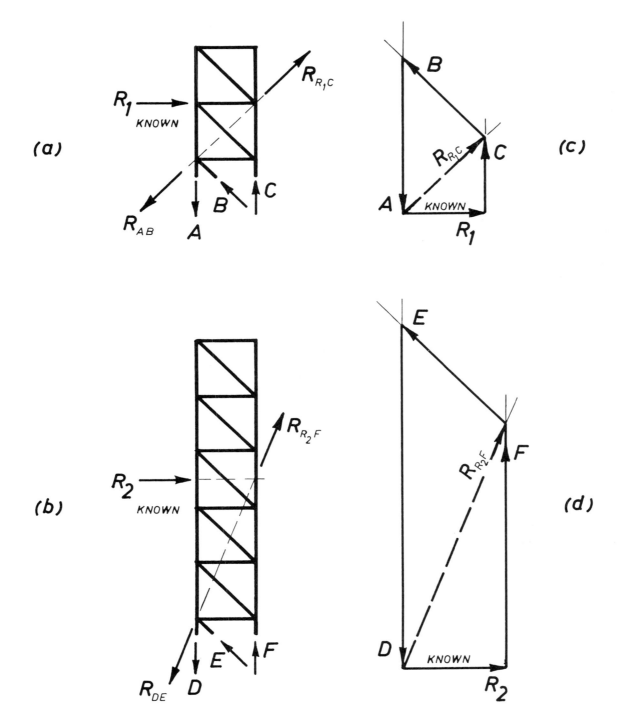

Figure 13.27 Solution–Tower Problem

PROBLEMS

All dimensions in millimetres unless otherwise specified.

P 13.1 Find the magnitude and direction of the resultant of the four concurrent coplanar vectors shown. Draw the space diagram, letter according to Bow's notation. Construct vector diagram to a scale of 1 mm = 1 Newton. Transfer resultant force to space diagram.

P 13.2 Determine the resultant of the three nonconcurrent coplanar vectors. Find the resultant of any two and add this resultant to the remaining vector. Use a scale of 1:100 for the space diagram and 1 mm = 10 Newtons for vectors. Show correct line of action of the resultant on the space diagram.

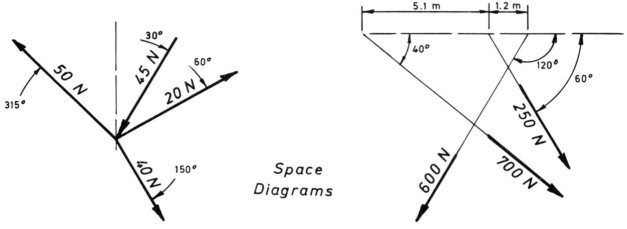

Figure P13.1

Figure P13.2

P 13.3 Three parallel forces act on a beam. Find the magnitude, direction, and line of action of the resultant. Draw the space diagram to a scale 1:50. Identify the forces using Bow's notation. Select a pole point and draw the vector diagram to a scale of 1 mm = 5 N. Use the funicular polygon to locate the line of action on the space diagram.

P 13.4 Three nonconcurrent coplanar vectors are not parallel to either the H or F plane. Find their resultant. Draw the space diagram to a scale of 1:50. Construct an auxiliary view in a plane to show the vectors in true length. Use Bow's notation in the auxiliary view. Draw the vector diagram taking directions from the auxiliary view. Use a scale 1 mm = 50 N. Project the resultant back to the space diagram and show it on the horizontal view.

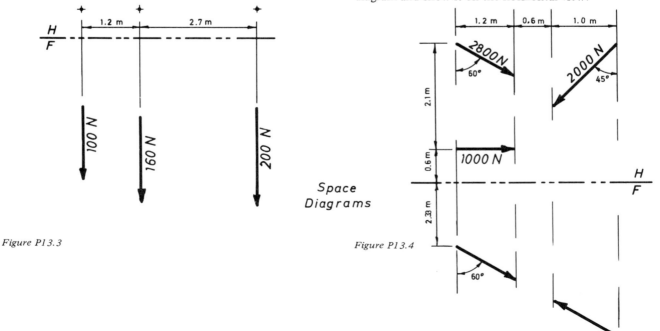

Figure P13.3

Figure P13.4

In the following two problems it is suggested that a sheet of tracing paper be taped over the space diagram. Trace significant details then remove the paper to complete the work.

P 13.5 Determine the resultant moment on the aircraft. Does the aircraft tend to nose up or nose down under the force conditions shown? Use a scale of 1:200 for the space diagram and 1 mm = 500 N for the force diagram.
T = 36500 N, L = 36000 N, F = 15000 N, D = 19000 N, W = 50000 N.

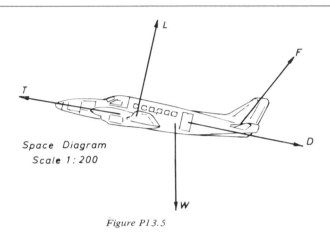

Figure P13.5

P 13.6 The latch beam shown has a fixed pin joint at A and a roller expansion pin joint at B. Determine the reactions at A and B. Use a scale of 1:20 for the space diagram and a scale of 1 mm = 100 N for the force diagram.

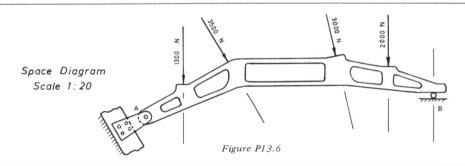

Figure P13.6

P 13.7 A sign weighing 1000 N is supported from a vertical wall by three rigid bars, pin-jointed at the ends. Determine the force in each bar and indicate whether it is in tension or compression. Scales: space diagram 1:50; vector diagram 1 mm = 20 N.

P 13.8 The double slider mechanism shown consists of a member AB, one metre long and two sliders moving in smooth slots. The resultant of the distributed weight of AB is 200 N and acts through its centre of gravity D, 0.4 m from B. For the position shown determine the reactions at A and B and the force P required to hold the assembly at rest. Scales: space diagram 1:10; vector diagram 1 mm = 2 N.

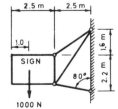

Figure P13.7

Space Diagrams

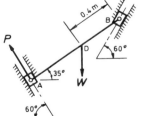

Figure P13.8

14. Vector Geometry – Noncoplanar Vectors

14.1 ADDITION OF NONCOPLANAR VECTORS

The general principal of vector addition as discussed in Section 13.1 is equally applicable to noncoplanar vectors. The situation is illustrated in Figure 14.1 which shows the addition of vectors A and B to form vector C using the parallelogram law. The same two vectors are shown in Figure 14.2 added according to the triangle law.

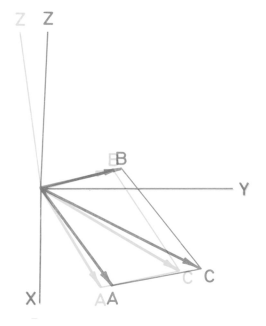

Figure 14.1 Parallelogram Addition of Vectors

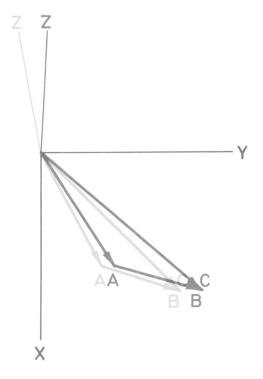

Figure 14.2 Triangle Addition of Vectors

14.2 NONCOPLANAR VECTORS AND ORTHOGRAPHIC VIEWS

Two or more orthographic views are required to describe completely noncoplanar vectors (vectors in three-dimensional space). That is, we require a Horizontal Projection and a Frontal Projection of both the space diagram *and* the vector diagram.

The fundamentals of vector addition, resolution, resultants, and equilibrium apply equally to these space vectors as they do to coplanar vectors. An important difference with space vectors is that the H and F projections do not necessarily show the true lengths. Therefore an additional operation of finding the true length is required.

14.3 RESULTANT OF NONCOPLANAR VECTORS

As an example consider the problem of determining the resultant of three noncoplanar vectors. The positions of the vectors are given in the space diagram together with their lines of action as shown in Figure 14.3 and 14.4. Remember that the length of the line of action on the space diagram bears no relation to the magnitude of a vector. In this example let the magnitudes of the three vectors be $F_1 = 100$ N, $F_2 = 90$ N, and $F_3 = 75$ N.

The resultant is obtained by adding the vectors in succession in a vector diagram. There will be two vector diagrams, one in the H plane and one in the F plane. The direction of a vector in the H projection plane of the vector diagram will be parallel to the line of action of that vector in the H plane of the space diagram. Similarly the direction of a vector in the F projection will be parallel to its line of action in the F plane of the space diagram.

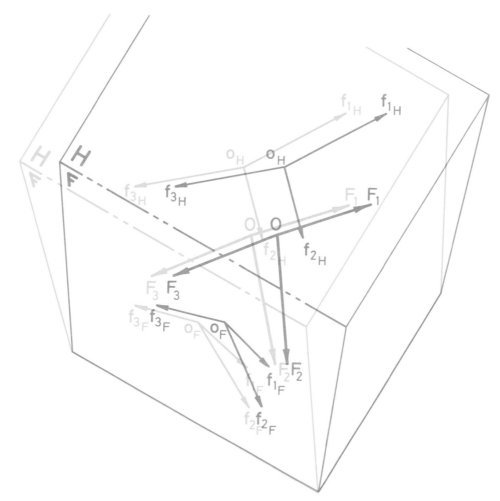

Figure 14.3 Three Noncoplanar Vectors—Space Diagram

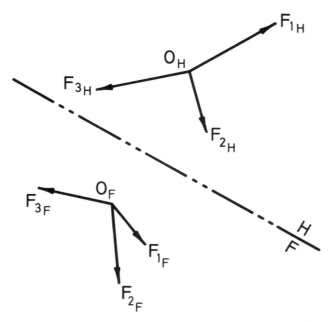

Figure 14.4 Projection of Three Noncoplanar Vectors—Space Diagram

To illustrate the procedures involved in drawing the vector diagram we will consider the force F_1 by itself and use a scale of 1 cm = 20 N. Since F_1 was given as 100 N its true length will be 5 cm. The H and F projections will not be 5 cm long, of course. To begin the vector diagram select some convenient starting point 0 in the H and F plane of the vector diagram as shown in Figure 14.5. Draw $o_H n_H$ parallel to $o_H f_{1_H}$ in the space diagram. Draw this line to any convenient length. Draw $o_F n_F$ parallel to $o_F f_{1_F}$ in the space diagram.

The next step is to obtain a true-length view of ON by either the rotation method or by projection on to an auxiliary plane. In this example an auxiliary projection plane 1, parallel to the F plane projection was selected and the true length $o_1 n_1$ was obtained by transferring the distances d_1 and d_2 from the H projection to locate point n_1.

On the true length $o_1 n_1$ scale off the required length of vector F (in this case 5 cm). This will locate the point f_{1_1} which may now be projected back to the H and F projections at f_{1_H} and f_{1_F} as shown.

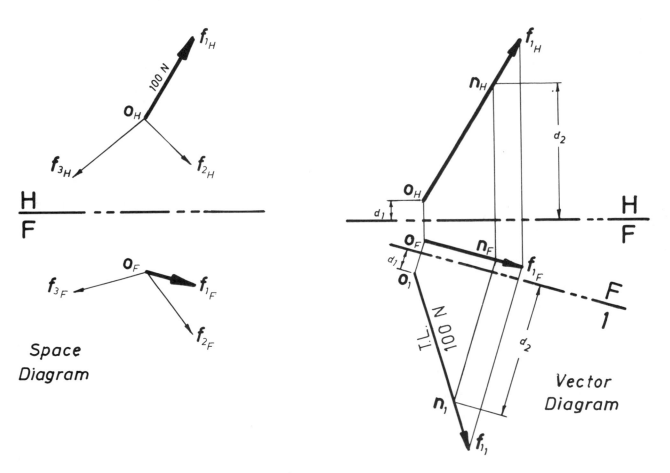

Figure 14.5 True Length by Auxiliary Plane Method

14.4 SHORTER METHOD OF DETERMINING TRUE LENGTH

It may be anticipated that the complete solution requires that the true lengths of the remaining vectors must also be obtained. With many vectors the procedure described above would complicate the drawing so it is preferable to use a shorter method based on the auxiliary plane method.

The shorter method is illustrated in Figure 14.6 in which we again consider the force F_1 only. Temporarily move the H/F folding line parallel to its initial position so that is passes through one end of the vector, in this case through the point o_H. Similarly the folding line F/1 has been placed so that it coincides with the frontal projection of the vector. That is, it is coincident with the line $o_F n_F$. The true length of ON is obtained by transferring the distance d_3 to locate the point n_1. On this true length measure the force $F_1 = 100$ N to the scale selected to find the point f_{1_1}. This point is then projected to the H and F views as shown.

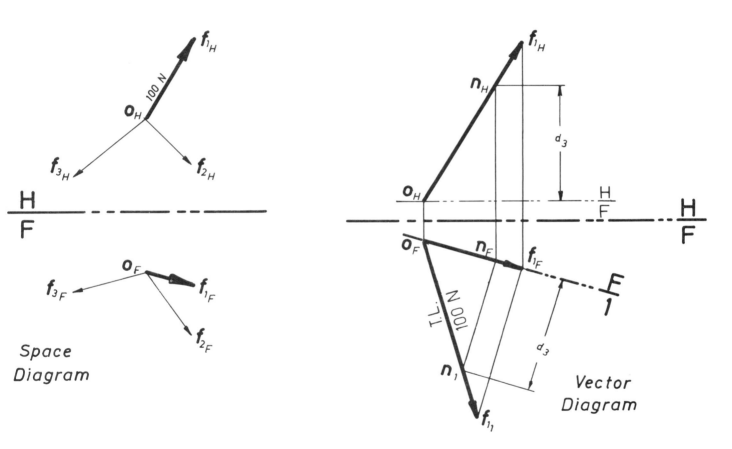

Figure 14.6 True Length by Shorter Method

Now that the correct lengths have been obtained for the H and F projections of the force vector F_1 we may proceed with the problem. The next step is to add the vector F_2 to F_1. Again we do not know the correct length of the H and F projections of the vector F_2. Therefore from the end of vector f_{1_H} add a vector (parallel to the space diagram direction of f_{2_H}) of any convenient length. Repeat the previous procedure to find the correct length of the projections. The procedure is shown in Figure 14.8.

Next add vector F_3 to F_2, finding the correct lengths as above.

The resultant is the vector R drawn from the origin 0 to the end of F_3. The true length of the resultant must be found so that it can be scaled to determine its magnitude.

The complete solution is shown in Figure 14.8 and indicates the resultant has a magnitude of 108 N. The general situation is shown in Figure 14.7.

Example 14.1 *Resultant of Noncoplanar Vectors*
A chair lift support tower has cable forces of 2000 N, 3000 N, and 3500 N applied as shown in the space diagram. Find the resultant.

Solution Assume a direction for the resultant R and show it on the horizontal projection. Number the spaces between the forces, including the resultant force, using Bow's notation. Proceed as detailed in section 14.4.

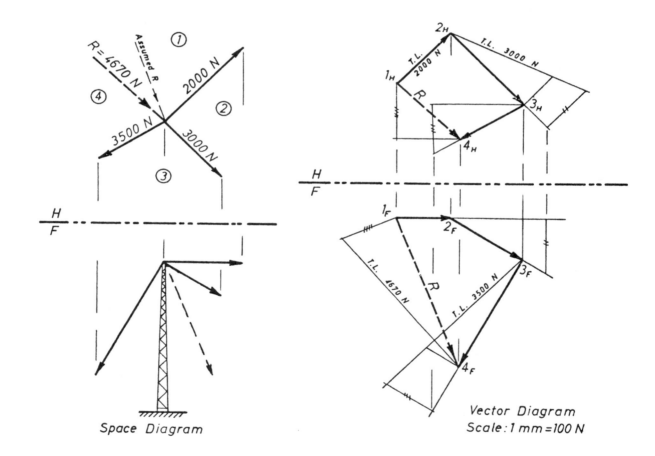

Space Diagram

Vector Diagram
Scale: 1 mm = 100 N

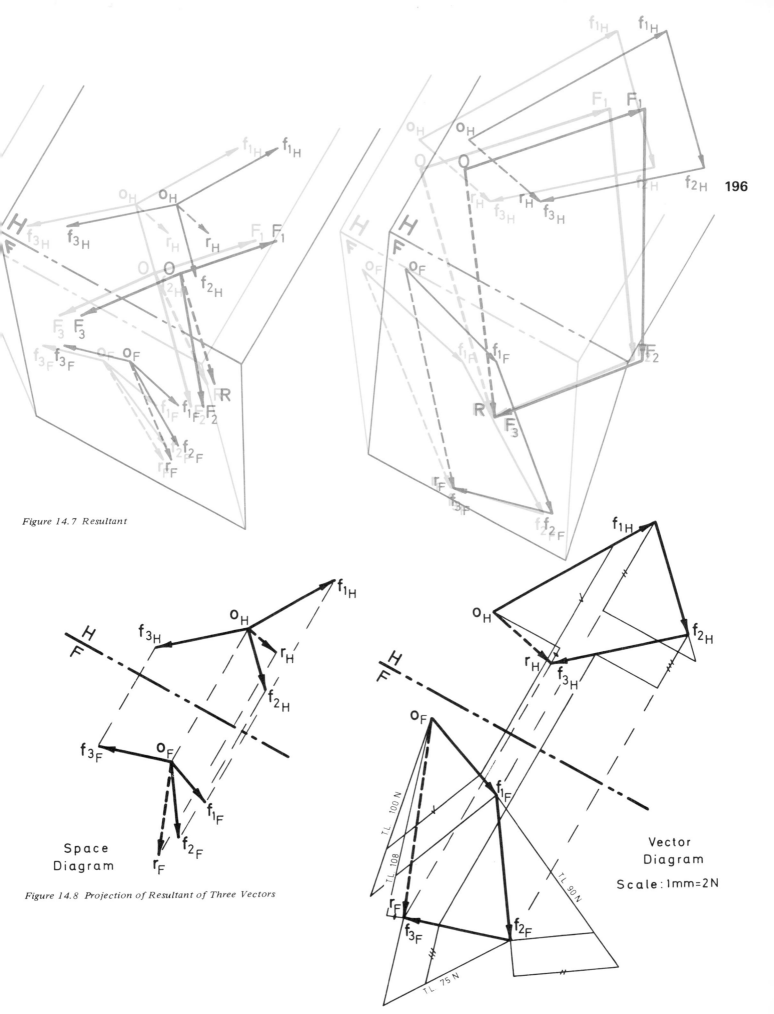

Figure 14.7 Resultant

Figure 14.8 Projection of Resultant of Three Vectors

Space Diagram

Vector Diagram

Scale: 1mm = 2N

14.5 EQUILIBRIUM OF NONCOPLANAR VECTORS

A body is in equilibrium if the sum of the forces on it are equal to zero, $\Sigma F = 0$, and if the sum of the moments are also equal to zero, $\Sigma M = 0$. The condition for $\Sigma F = 0$ is that the force polygon will close. For *concurrent* forces $\Sigma M = 0$ has no meaning as there is no moment arm. Generally we are faced with the problem of finding the forces required to keep a system in equilibrium.

In any one view of a system of *concurrent* noncoplanar forces the magnitudes of only two unknown vectors may be found. This is because only two dimensions are available in any one view. For example the x and z dimensions are found in the H plane projection. Therefore the closure of a force polygon in this projection will satisfy the two equations $\Sigma F_x = 0$ and $\Sigma F_z = 0$ and with two equations we can solve for two unknowns. However by using an adjacent view it is possible to solve for *three* unknown magnitudes as we have one more equation, $\Sigma F_y = 0$.

When it is necessary to determine three unknown magnitudes the views must be selected so that we obtain either:
1) a point view of one of the known or unknown forces, or
2) an edge view of the plane containing the lines of action of two of the forces.

A view such as the above will effectively reduce the number of unknowns in that particular view so that a solution may be obtained.

Point View Example

As an illustration of the use of a point view consider the problem of determining the forces in the legs of a tripod structure supporting a load which is pushing vertically downwards. The general situation is shown in Figure 14.9. In this problem the maximum permissible load in leg OA is 1500 N and we are required to find the load the tripod will support and also the loads in the legs OB and OC.

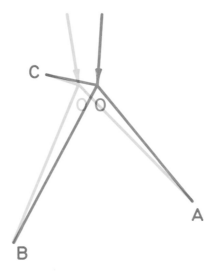

Figure 14.9 Vertically Loaded Tripod

The first step is to draw the H and F projections of the space diagram as shown in Figure 14.11. Note that the H view shows the vertical force as a point view. This is necessary as we have three unknowns and if we reduce these to two in this view we can obtain a solution as the force triangle can be closed. Use Bow's notation in the frontal F projection as it is the view which shows all four lines of action.

Draw an H/F line for the vector diagram parallel to the H/F line for the space diagram and select a convenient starting point, 1, for the known load 12. The load 41 is a vertical load therefore the horizontal projection of point 4 will be coincident with point 1.

Start the vector polygons with the known load 12 and draw it in both projections parallel to their respective space diagram lines of action and to the correct length as detailed earlier in Section 14.4. This locates points 2_H and 2_F. Add vector 23 to 12 in the H view and then vector 34 to complete the vector triangle in the H view. The frontal projection of 23 is added to vector 12 and is terminated at the point 3_F projected from 3_H. The vector 34 will complete the polygon. The *true* lengths of vectors 23, 31 and 41 must be obtained and scaled to provide the required answers.

After the polygons are completed the arrowheads are placed on the vectors 12, 23, etc. in both views, the tail being at the first number and the head at the last. When these directions are transferred to the space diagram we are able to determine which members are in tension and which are in compression. In this case all the arrowheads point towards the point 0 so that all members are in compression.

It may be noted that in starting to draw the force polygon in Figure 14.11, force 12 was assumed to have a direction upwards to the left. This assumption proved to be correct as the force triangle could be closed. If, however, we had assumed the direction of 12 to be downwards and to the right, the vector triangle could not be closed. Therefore this latter assumption would have been incorrect.

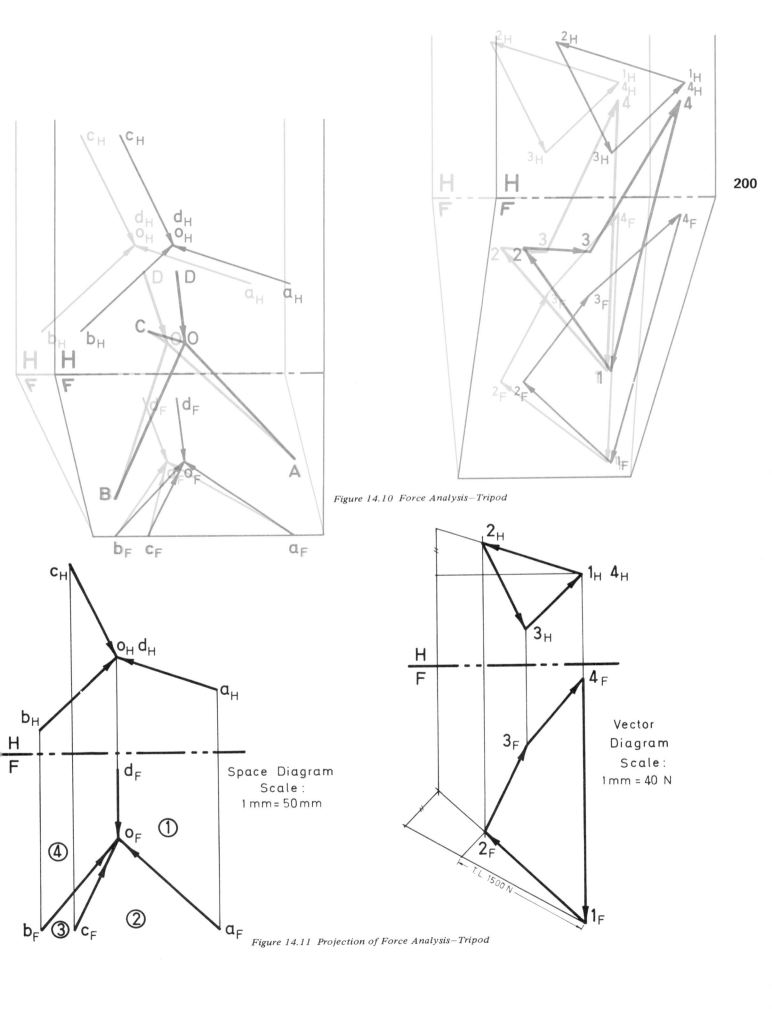

Figure 14.10 Force Analysis—Tripod

Figure 14.11 Projection of Force Analysis—Tripod

14.6 TRIAL SOLUTION METHOD

In the above method if the vector for which the point view was taken is a known quantity, then the procedure will not yield a solution. However we can solve this problem by assuming a trial solution.

Consider the same tripod but this time determine the loads in the legs given that the vertical load is 6000 N. Draw the space diagram and select a point 1 in the vector diagram as before.

Since we have no value for the force 12, assume a convenient trial length to fix the trial points $2'_H$ and $2'_F$ as shown in Figure 14.12. Proceed as before to complete the trial polygons in each view. Measure the length of the trial vector $4'_F 1_F$ and equate to the load of 6000 N to obtain the vector diagram scale. This scale may not be convenient so we could choose a more convenient scale and redraw the polygons to the new convenient scale.

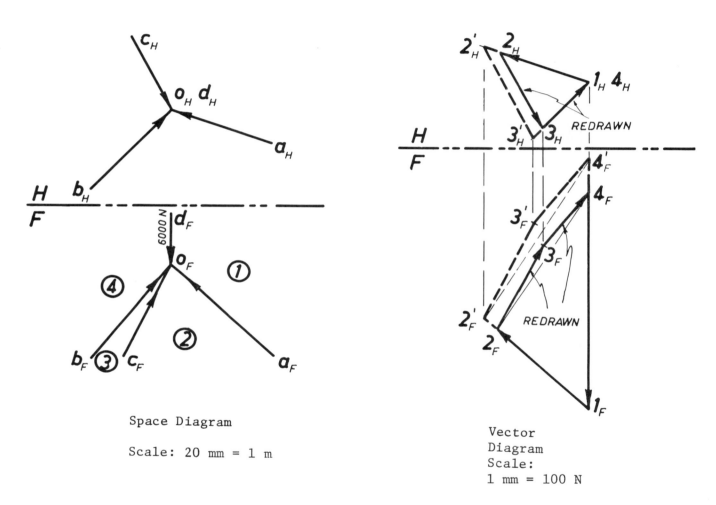

Figure 14.12 Trial Solution Method

14.7 EDGE-VIEW METHOD

In this method a view of the space diagram is obtained such that the plane formed by two of the members is seen in an edge view. The vector diagram is constructed from this edge-view projection and not from the original H and F plane projections.

As an example of the edge-view method consider a tripod loaded in a non-vertical direction as shown in Figure 14.13. (We could obtain a point view of one of the members by the use of a double auxiliary view however it is quicker to use the edge-view method.)

In the edge-view method the projection of two of the legs will coincide. For example if we look in the direction shown in Figure 14.13 the plane formed by legs OA and OC will appear as an edge view.

In this example the H and F projections of the tripod legs are given in Figure 14.15. We are required to find the forces in the legs of the tripod when an applied load of 4000 N acts in the oblique direction, DO, shown.

To obtain an edge view of the plane OAC we can take a point view of the horizontal line AC which lies in the plane. The auxiliary projection in plane 1 is perpendicular to the true length of AC. This view will provide the required point view of AC and the edge view of plane OAC.

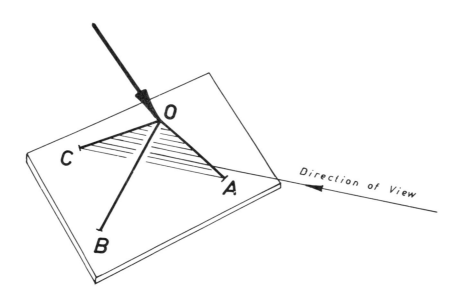

Figure 14.13 Direction of View

The view 1–1 is a vertical projection of the space diagram and will be used in conjunction with the H projection to draw the vector diagram. Note that in the vector diagram we use a folding line parallel to the H/1 line and not parallel to the H/F line as in previous examples. The general situation is shown in Figure 14.14 with the orthographic projections shown in Figure 14.15.

When the vector diagram is completed and the arrow heads placed in sequence around the polygon they are transferred to the space diagram. From the space diagram we can see that leg OA is in compression and legs OB and OC are in tension.

When using the edge-view method it is essential that the two vectors in the plane selected to appear in edge view must be adjacent to each other. For example, in the above situation we cannot use plane OCB because the vector DO (12) comes between OB (41) and OC (23).

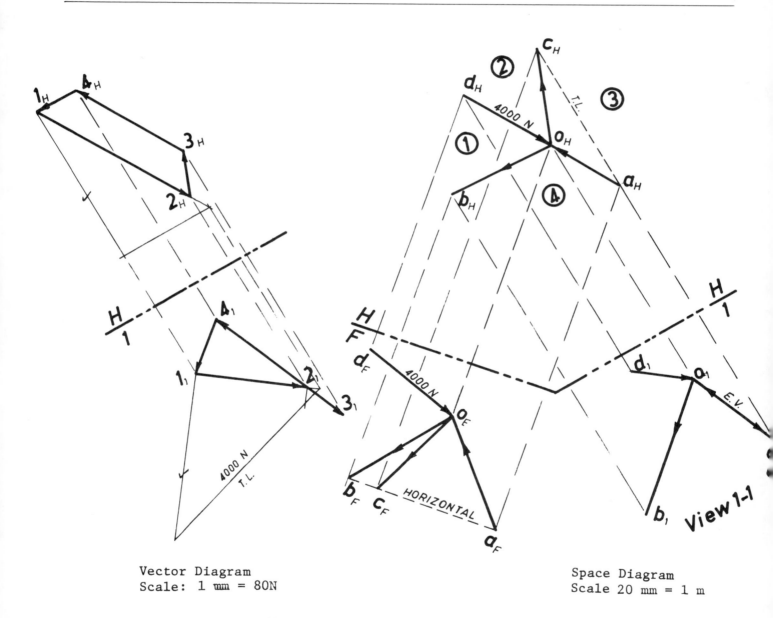

Figure 14.15 Orthographic Projection – Tripod Problem

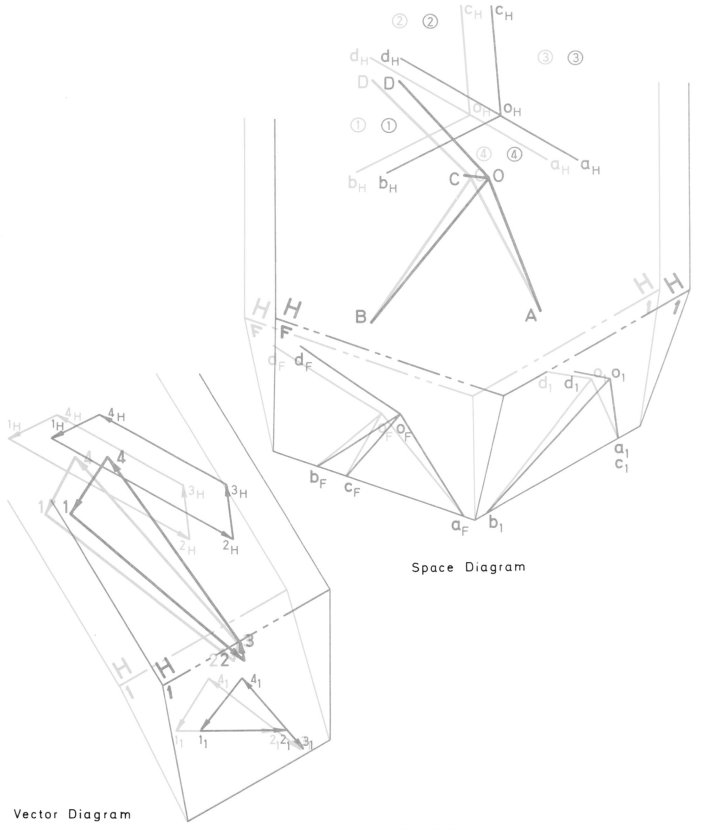

Figure 14.14 Edge-View Solution — Tripod Problem

PROBLEMS

All dimensions in millimetres unless otherwise specified.

P 14.1 Determine the resultant of the two concurrent noncoplanar forces shown. Report the magnitude and show the resultant on the space diagram. Scales: space diagram 1:1; vector diagram 1 mm = 1 Newton. OA = 40 N, OB = 40 N.

P 14.2 A support tower for a ski-lift has three cable forces of 2000 N, 3800 N, and 4400 N acting as shown. Find the resultant of these forces and show it on the space diagram. What is the vertical force in the tower itself?

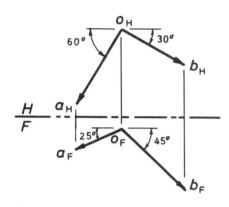

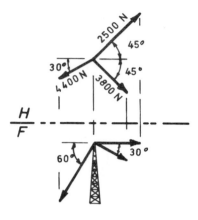

Space Diagrams

Figure P14.1

Figure P14.2

P 14.3 During an underwater salvage operation a diver's tool cage was constructed as shown because of certain clearance problems. If the total load in the vertical lifting cable is 2500 N find the forces in each of the three support cables. Use trial solution method.

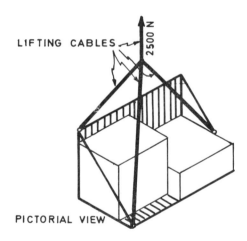

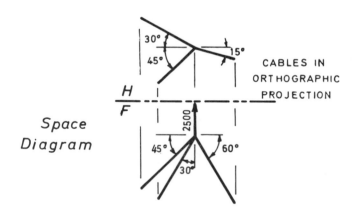

Space Diagram

Figure P14.3

P 14.4 Explosive charges producing a known force have been used as self-contained light-weight power sources. Examples of use are pilot's ejection seats and emergency cable cutters for helicopter lifting slings. In this example an explosive charge is used to actuate a device to pull out three release pins in an emergency exit hatch. The charge is ignited in the cylinder, forcing the piston to move out with a force of 3800 N, to pull the pins. Take a point view of the unknown force in cable OB by using an auxiliary view. Find the forces in each cable. Vector scale 1 mm = 50 N.

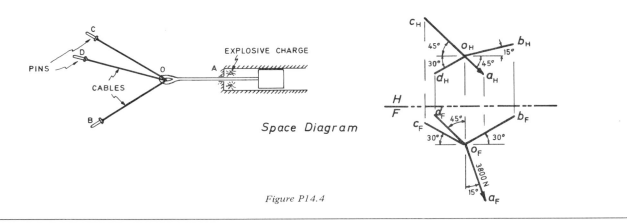

Figure P14.4

P 14.5 Solve problem P 14.4 by taking an edge view of plane OBC.

P 14.6 Determine the magnitude and direction, that is the bearing, slope, and sense (direction of the arrowhead), of the resultant of the three orthogonal vectors shown.

Figure P14.6

P 14.8 Find the magnitude of the forces in each leg of the tripod shown. State whether each force is tensile or compressive. Force DO = 5000 N.

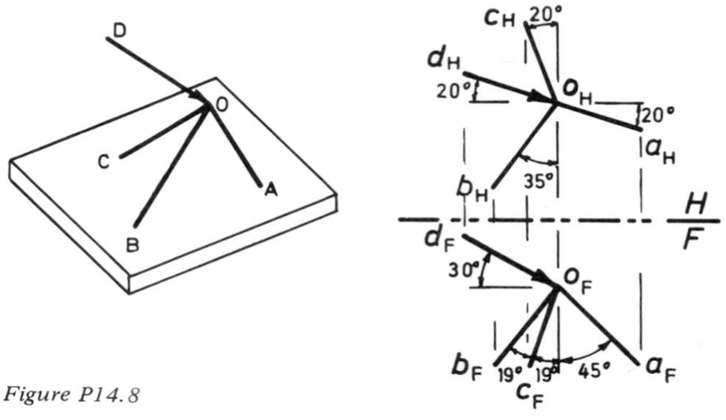

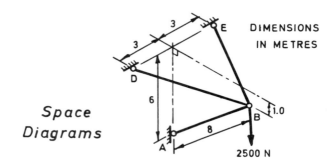

Figure P14.8

P 14.7 Determine the forces in each member of the hanging cable system shown.

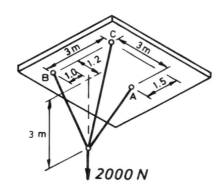

Figure P14.7

P 14.9 A load of 5000 N is suspended from a vertical wall by the cable supported strut AB. Determine the force in each cable BD and BE and the reaction force at A.

DIMENSIONS IN METRES

Space Diagrams

Figure P14.9

The following problems involve coplanar vectors (Chapter 13 applies).

P 14.10 Use the method of joints to determine the forces in members AC, BC, and BD of the truss loaded as shown. A is a pin joint and B is a roller joint. Determine the reactions at A and B.

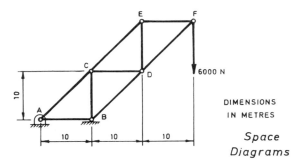

Figure P14.10

P 14.11 The load a crane may lift depends on the position of the boom, that is on the vertical lift and horizontal reach. If the maximum force in the boom is to be 10000 N, determine the permissible load W. Plot a graph of load W against angle of inclination θ (in increments of 15°) as the boom goes from the horizontal position to the $\theta = 75°$ position. AB = 4.2 m. Use a space diagram scale of 1:50.

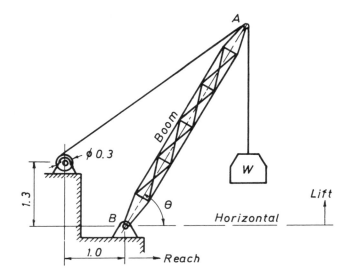

Figure P14.11

15. Kinematics

Kinematics may be defined as the geometry of motion. It includes the displacement, velocity, and acceleration of points or rigid bodies. This chapter will consider only the displacement and velocity for a few mechanisms.

15.1 DISPLACEMENT

The displacement of a point on a simple mechanism can be determined by inspection. For example, the motion of the crank on the slider crank mechanism shown in Figure 15.1 is circular with a diameter equal to the stroke. The motion of the slider or piston is back and forth (rectlinear).

In the case of more complex mechanisms a graphical layout may be necessary to analyse the motion. For example the path traced out by the end of the rod B in the film feed mechanism shown in Figure 15.2 was obtained by laying out the mechanism at 15° intervals and finding the position of B at each interval. A smooth curve joining these consecutive positions is the locus of point B during one cycle of operation.

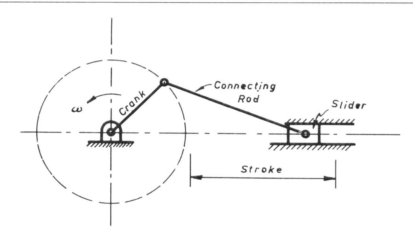

Figure 15.1 Slider Crank Mechanism

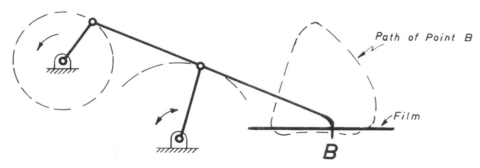

Figure 15.2 Path of a Point on a Mechanism

15.2 POSITION SYNTHESIS OF A LINK

Synthesis may be defined as the selection of links of a mechanism in order to perform a specified motion. The individual parts of a mechanism are called links. A common design requirement is to have some part take up certain definite positions. The simplest example is a door which is hinged so that it takes up the open or closed position. A more complicated example is shown in Figure 15.3 where the link BC is to move from position B_1C_1 to position B_2C_2. Point B goes from B_1 to B_2 and the right bisector of B_1B_2 is the locus of centres of circles passing through B_1 and B_2. Therefore, the solution is accomplished by selecting any convenient centres O_1 and O_2 on the perpendicular bisectors of B_1B_2 and C_1C_2. It can be seen that an infinite number of solutions are possible. This is known as two-position synthesis.

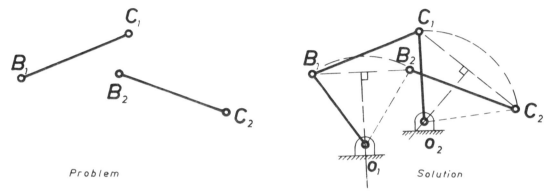

Figure 15.3 Two-Position Synthesis

15.3 ILLUSTRATIVE EXAMPLE OF PLANAR TWO-POSITION SYNTHESIS

It is required to move the machine part shown in Figure 15.4(a) from position 1 to position 2 and back again. The motion is all in the plane of the paper (planar).

First we will make a cradle to hold the part and to which we can attach pivots for the links at points A_1 and B_1 as shown in Figure 15.4(b). The problem is to design a mechanism to move the cradle from position A_1B_1 to A_2B_2.

Join A_1A_2 and B_1B_2. Construct the right bisectors of lines A_1A_2 and B_1B_2. We can choose any convenient points C and D on these bisectors to form the linkage CABD. This linkage will perform the function of moving the part from position 1 to position 2 and return as shown in Figure 15.4(b).

In the case of two-position synthesis we can simplify the mechanism by considering the intersection of the bisectors, point E, Figure 15.4(c). If this point is chosen as a hinge point then a single rotating arm can be used to replace links CA and DB. Point E represents the axis of rotation of the arm.

Figure 15.4(a) Location of Required Positions

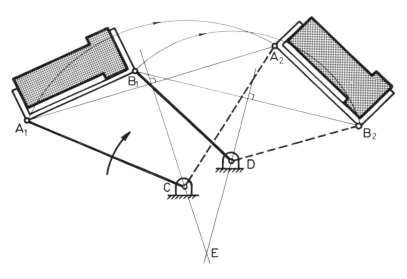

Figure 15.4(b) Geometrical Construction to Obtain Linkage

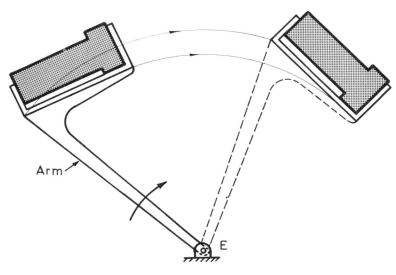

Figure 15.4(c) Single Arm to Replace Linkage

15.4 ILLUSTRATIVE EXAMPLE OF SPATIAL TWO-POSITION SYNTHESIS

In this example it is required to move an object from one position in space to another. In this case we wish to move the object represented by the line A_1B_1 to the position A_2B_2 as shown in Figure 15.5.

This can be done by finding an axis about which to rotate the line in an analogous manner to the previous problem. Since A goes from A_1 to A_2 the perpendicular bisector of the line A_1A_2 is the locus of the centres of circles passing through A_1 and A_2. However since this is a spatial problem we require the perpendicular bisecting plane of the line A_1A_2. Similarly we require the perpendicular bisecting plane of the line B_1B_2. These two bisecting planes will intersect to form an axis about which the line AB may be revolved. We may choose any point on this axis as the location of an arm to revolve the object line as required. This is illustrated in Figure 15.7.

The orthographic projection is shown to a reduced scale in Figure 15.6. The procedure will be explained with reference to the orthographic projection.

Procedure

Given the horizontal and frontal projections of A_1, A_2, B_1, and B_2 find the axis required to revolve A_1B_1 to position A_2B_2.

Join A_1A_2 and B_1B_2 and obtain true length views of these lines. The frontal projection plane has been chosen so that $b_{1_F} b_{2_F}$ will be a true length. The true length of A_1A_2 is shown as $a_{1_2} a_{2_2}$ in auxiliary plane 2.

Obtain a point view of B_1B_2 in auxiliary plane 1. The right bisecting plane will appear true shape in this view and may be constructed to a convenient size as shown by $c_1d_1e_1f_1$.

Similarly obtain a point view of A_1A_2 in plane 3. Construct the right bisecting plane $g_3h_3j_3k_3$.

To find the intersection, XY, of the two planes CDEF and GHJK project one plane back to a view where the other plane is seen in edge view. In this case plane GHJK was projected from view 3 back to the frontal view where plane CDEF appears as a line. The intersection $x_F y_F$ is found in this frontal view.

The required axis of rotation is XY.

Additional details of the general problem are shown in Figure 15.7 which is drawn to a reduced size.

214

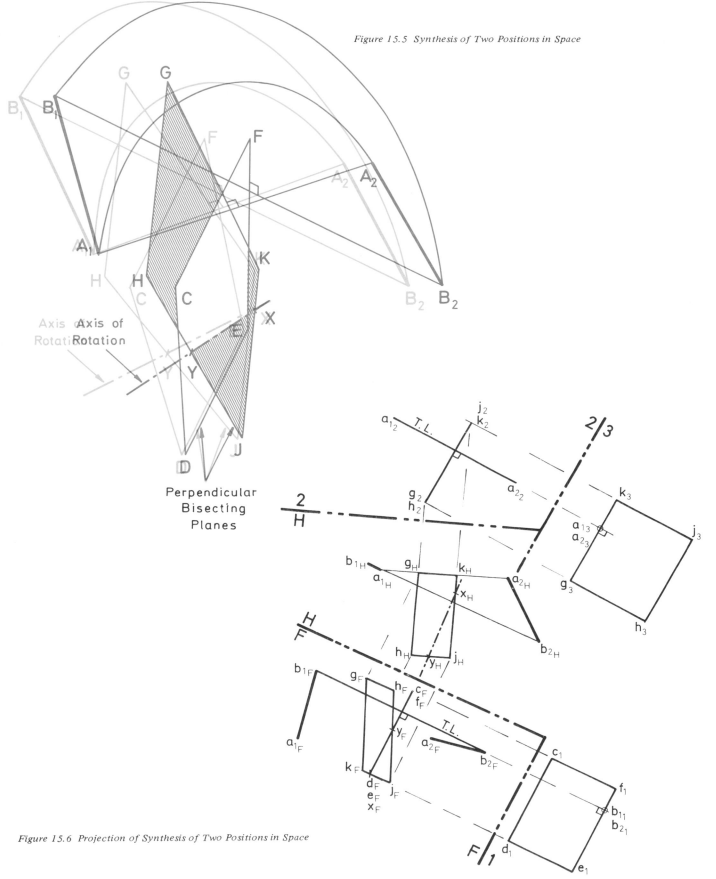

Figure 15.5 Synthesis of Two Positions in Space

Figure 15.6 Projection of Synthesis of Two Positions in Space

The drawings on the opposite page show some additional views of the solution to this spatial two-position problem. In order to get the views within the page they have been reduced in size. In addition they have been rotated from the position shown in Figure 15.5 for purposes of illustration only.

The top left drawing, Figure 15.7(a) shows the arcs of rotation that each point will move along. This drawing has been rotated 45° in a clockwise direction. The points L and M are chosen so that $\angle B_1LY = \angle B_2LY = 90°$ and $\angle A_1MY = \angle A_2MY = 90°$. Point L or M may be used to find the angle of rotation of the line AB about the axis when it moves from position 1 to position 2. The angle of rotation is $\angle B_1LB_2 = A_1MA_2$. The points L may be found in the orthographic projection, Figure 15.6, in view 3 by dropping a perpendicular from $a_{1_3}a_{2_3}$ to x_3y_3. Similarly point M may be found in view 1. The angle of rotation must be measured in a true size view of the plane A_1MA_2 or plane B_1LB_2 (not shown in Figure 15.6).

The top right drawing Figure 15.7(b) shows an arm to rotate the line AB around the axis. This arm can be located at any convenient position along the axis XY. This drawing has also been rotated 45° clockwise from the position in Figure 15.5.

The lower drawing, Figure 15.7(c) shows the formation of the hyperboloid of revolution obtained by line AB as it rotates about the axis. Refer to Chapter 11 where it was shown that a hyperbolic surface is obtained when a line rotates about another line. This drawing has been revolved 90° in a clockwise direction from the position shown in Figure 15.5.

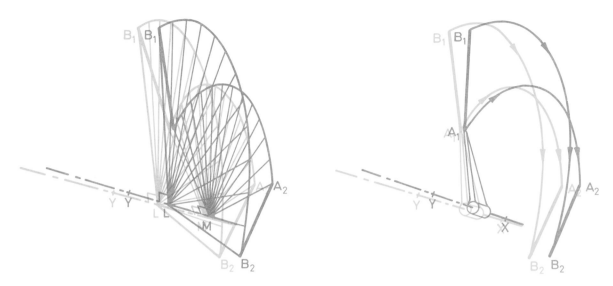

Figure 15.7(a) Arcs of Rotation

Figure 15.7(b) Rotating Arm

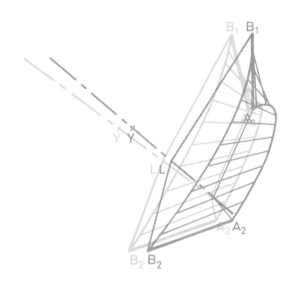

Figure 15.7(c) Hyperboloid of Revolution

15.5 THREE-POSITION SYNTHESIS

When three positions of a link are required the problem is a little more complex. Consider the case where a telescope support tray on a space craft has to take up three positions. The three positions might be stowed, left, and right as indicated in Figure 15.8. The perpendicular bisectors result in only one possible position for the hinge points $O_3 O_4$, if we are restricted to making connections to the link at the end points D and E. If no such restrictions exist there are an infinite number of possible solutions.

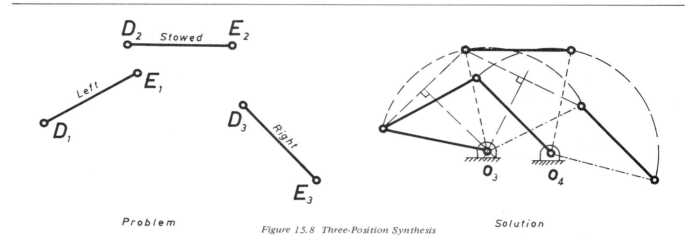

Figure 15.8 Three-Position Synthesis

15.6 RELATIVE VELOCITY

Relative velocity between two points is the apparent velocity of one point as seen by an observer moving with the other point. If we call one point A and the other B the relative velocity of B with respect to A may be written as the vector equation:

$$V_{BA} = V_B - V_A$$

where V_{BA} = velocity of B with respect to A
V_B = absolute velocity of B
V_A = absolute velocity of A.

The concept of relative velocity is useful in the analysis of mechanisms. Consider the slider crank mechanism where we require the piston velocity knowing the speed of rotation of the crank, Figure 15.9.

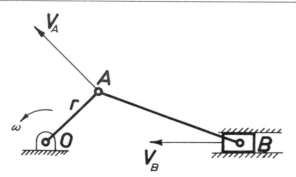

Figure 15.9 Velocities in a Slider Crank Mechanism

Since we want to find V_B write the equation as
$$V_B = V_A + V_{BA}$$
The velocity of B is known in direction as it is constrained to move horizontally. The velocity of A is known in magnitude and direction. The magnitude to $r\omega$ mm/sec. and the direction is at right angles to the link OA. The direction of the velocity of point B with respect to A (V_{BA}) is at right angles to the link AB. This is because the link is rigid and to an observer at A the only motion B can have relative to the observer is at right angles to the line AB.

The vector equation can be solved graphically as shown in Figure 15.10. The point 0' is the ground point of zero velocity. Draw V_A to scale from 0' to find A'. Draw V_B in horizontal direction. Through A' draw V_{BA} to meet V_B at B'.

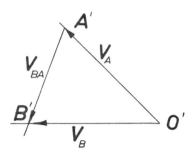

Figure 15.10 Vector Diagram of Relative Velocity Equation

15.7 SPACE MECHANISM—DISPLACEMENT

All actual mechanisms are solid, i.e., three-dimensional but many can be represented by two-dimensional links and coplanar velocity vectors as indicated in the previous sections. On the other hand some mechanisms operate in three dimensions, such as the double slider mechanism of Figure 15.11. This type of mechanism may be called a *space mechanism*.

The displacement of one part of a mechanism given the motion of another part can be determined by a layout using the principles of descriptive geometry. As an example suppose the height of A is 30 mm and the length of the link AB is 95 mm in Figure 15.11. The length of CB may be determined from the right angled triangle ACB. With CB known the layout of Figure 15.12 can be made. If now the link A is moved down 1 cm, say, a new layout will be required to find the new position of point B. When point A moves 1 cm point B does not necessarily move 1 cm. The actual movement of B depends on the initial position of the sliders.

Example 15.1 *Offset Slider Crank — Displacement and Velocity*

For the given mechanism determine the stroke length and the velocity of the slider at the position shown. The crank rotates at 30 radians per second.

Solution The two extreme positions of travel will occur (1) when the crank and connecting rod are fully extended and (2) when they are folded back over each other. Draw these two positions to scale and measure the stroke.

Choose an origin for vector diagram. Using $V_B = V_A + V_{BA}$ calculate $V_A = r\omega$. Draw this vector at right angles to 0A. Add V_{BA} (perpendicular to BA) to meet horizontal velocity line of V_B at B'. Scale off 0'B'.

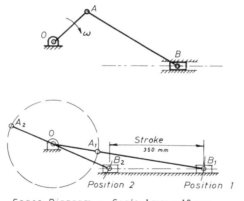

Space Diagram - Scale: 1 mm = 10 mm

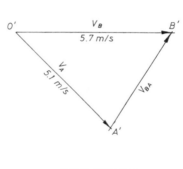

Vector Diagram
Scale: 1 mm = 0.1 m/sec.

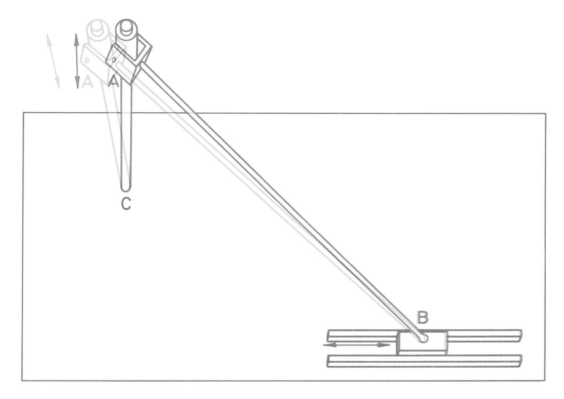

Figure 15.11 Double Slider Mechanism

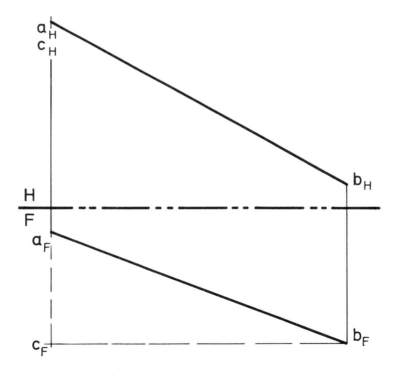

Figure 15.12 Projection of Double Slider Mechanism

15.8 SPACE MECHANISM—VELOCITY

The relative velocity equation may be used to determine the velocity of one point on a space mechanism, given the velocity of another point. In this case, however, the vector diagram involved must be drawn in two projections. In addition the velocity of one end of a rigid link relative to the other end is perpendicular to the true length of the link. This means that usually an auxiliary plane is required to determine the true length view.

Consider the problem of finding the velocity of slider B when slider A is given a downward velocity of 50 mm/sec, Figure 15.13. The relative velocity equation may be written as

$$V_B = V_A + V_{BA}$$

The velocity of B with respect to A, V_{BA}, is at right angles to the true length of the link BA. The true length of BA will be found in plane 1 which is parallel to the H plane projection $a_H b_H$, Figure 15.14.

Start the vector polygon at a convenient origin o'_1 and draw the velocity of A_3 V_A, to a suitable scale, in a vertically downward direction to locate point a'_1. In the H plane projection o'_H and a'_H will coincide because A' is directly below o'.

Through the point a'_1 draw a line perpendicular to $a_1 b_1$. This represents the direction of the velocity vector V_{BA}. Through o'_1, draw a horizontal line representing the velocity of B, to close the vector triangle at b'_1. Draw $o'_H b'_H$ in the direction of the motion of slider B to locate b'_H. This is a true length and may be scaled.

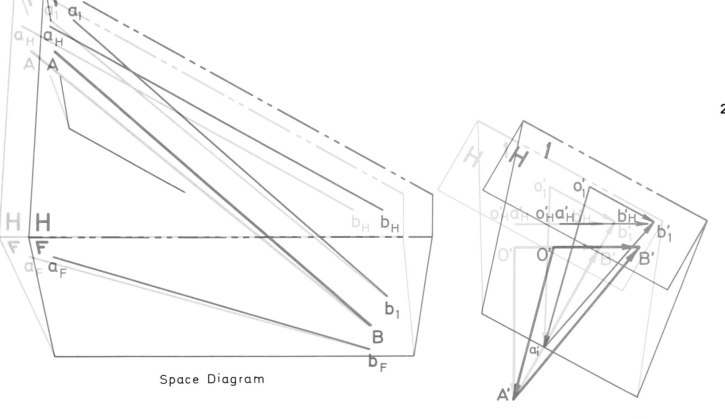

Figure 15.13 Velocity of Double Slider Mechanism

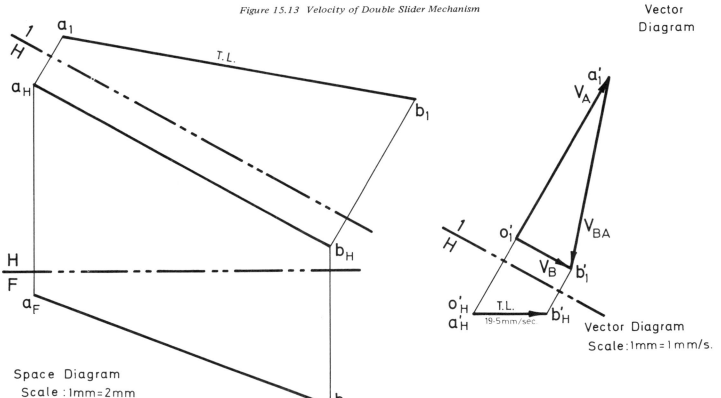

Figure 15.14 Projection of Double Slider Velocity

PROBLEMS

All dimensions in millimetres unless otherwise specified.

P 15.1 The rod BA slides freely through the block which is pivoted at C. Plot the path of point A as the crank OB makes one revolution. Divide the crank circle into 24 equal 15° parts and for each position of B locate point A on the rigid link BA.

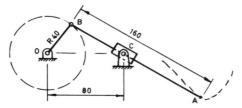

Figure P15.1

P 15.2 The Watts mechanism is designed to produce a straight line by a recording pen held at point P as OA rotates. Lay out successive positions of OA on either side of the position shown to determine the length of the straight line portion of the path of P.

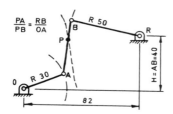

Figure P15.2

P 15.3 Design a planar four-bar mechanism to move the hatch from position A_1B_1 to A_2B_2, (i.e., lay out the link centre lines and pivot points). Because of space constraints no part of the linkage may be placed in the cross-hatched area and the complete linkage must lie within the circular section of the vehicle.

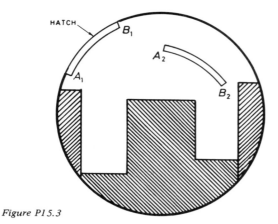

Figure P15.3

P 15.4 Design a planar four-bar mechanism to move the block from position 1 to position 2.

Figure P15.4

P 15.5 Design a spatial mechanism to transfer the link AB from position A_1B_1 to position A_2B_2 by rotation.

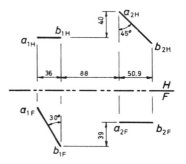

Figure P15.5

P 15.6 Design a mechanism to move the probe from position 1 to position 2 to position 3.

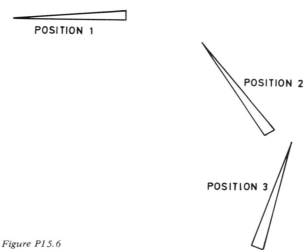

Figure P15.6

P 15.7 The link OA of the offset slider crank mechanism rotates anticlockwise. The constant velocity of point A is $r\omega$ (ω = 10 rad/sec). Slider B is constrained to move along the line XY. OA = 80 mm, AB = 200 mm, offset = 50 mm.
a) Draw the mechanism in 12 successive positions, $\theta = 0°, 30°, 60°$, etc. Use a scale 1:2.
b) Draw the velocity diagrams for each 30° in the place indicated. Scale 1 mm = 20 mm/sec.
c) Draw a graph of velocity of B versus θ. Use velocity scale of (b) and 25 mm = $\pi/2$ for θ.

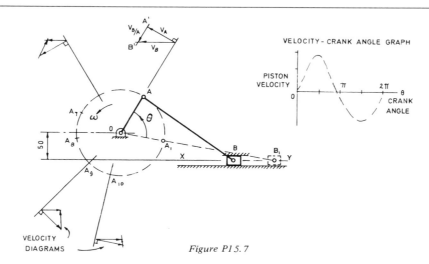

Figure P15.7

P 15.8 A four-link mechanism is shown in Figure P 15.8(a). (The fourth link is considered to be the ground.) If ω = 20 radians per second clockwise, what is the velocity of point B? What is the angular velocity of link BC in radians per second? For the position shown in Figure P 15.8(b) what is the velocity of point B?

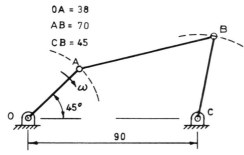

Figure P15.8(a)

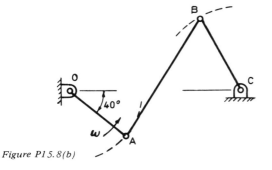

Figure P15.8(b)

P 15.9 The mechanism for a front end loader is shown in Figure P 15.9. The specifications call for the boom to elevate through 74° from the position shown. Find the piston travel required for the boom cylinder X to produce this elevation. At the upper position the bottom of the bucket is to slope downward at 50° to dump the contents. What piston travel is required for bucket cylinder Y to accomplish this operation? Use a scale of 1:10.

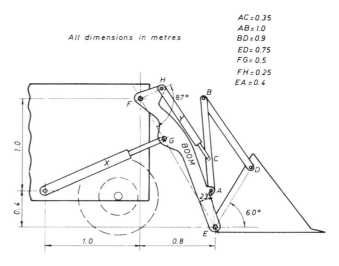

Figure P15.9

P 15.10 Slider Crank Space Mechanism—Displacement

The isometric view, Figure P 15.10(a), shows a slider crank mechanism operating in space rather than in a plane. The crank is on a disc, the crank radius being 34 mm.

The significant dimensions are shown on the orthographic views of Figure P 15.10(c). The crank in these views is in a general position.

You are required to find the displacement of the slider as the crank rotates from one horizontal position to the other horizontal position (180° from the first).

Procedure Lay out the significant features in the position indicated in Figure P 15.10(b). Draw the H and F projections. Draw the profile projection of the crank only. Use a scale of 1:2.

Draw auxiliary views as necessary to fit the 150 mm long connecting rod in the correct location for both positions.

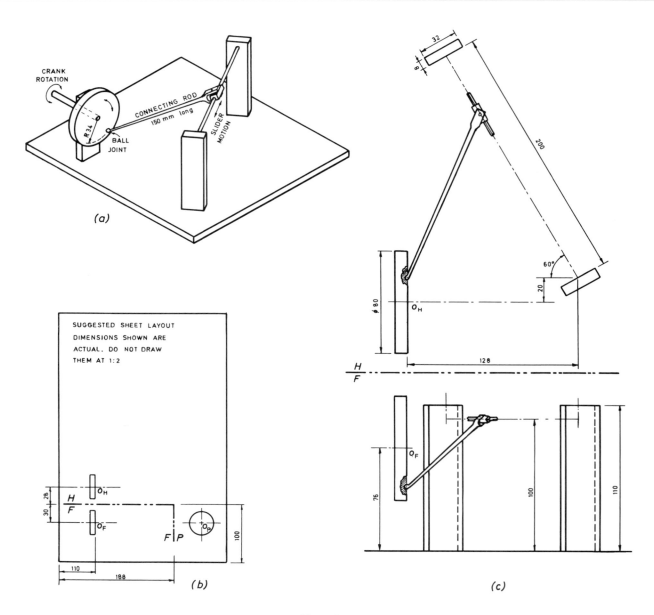

Figure P15.10

P 15.11 A spatial slider crank mechanism is shown in Figure P 15.11. How far will the slider move if the crank OA rotates from the position shown, through 45° in the direction shown? A and B are ball joints.

P 15.12 If the angular velocity, ω, of OA is 20 rad/sec at the point shown, what is the velocity of the slider B? Figure P 15.12.

P 15.13 A spatial four-link mechanism is shown pictorially in Figure P 15.13(a) and in orthographic projection in Figure P 15.13(b). If crank OA rotates through 30° in the direction shown, from the position shown, through what angle does link MN rotate?
Note: In making the orthographic layout you will have to make use of the true length of link AN. The rotation method of true length is convenient.

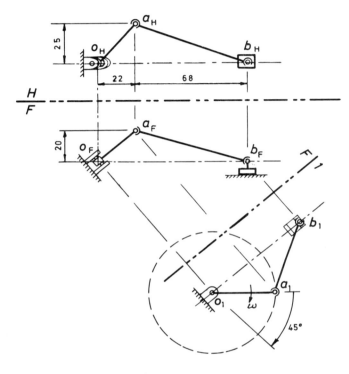

Figures P15.11 and P15.12

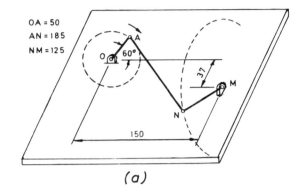

Figure P15.13

P 15.14 If the angular velocity of OA is 20 rad/sec what is the velocity of point N and the angular velocity of NM at the position shown.

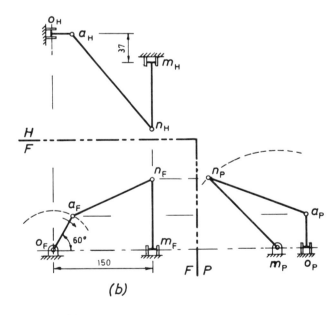

Figure P15.14

16. Graphical Calculus

Two useful concepts of graphical calculus, differentiation and integration, will be developed in this section. Differentiation is commonly associated with the slope of a function and integration is associated with the area under a curve. More precisely differentiation involves the rate of change of variables and integration involves the inverse process to differentiation.

The standard mathematical techniques of differentiating and integrating are used when convenient, however it is sometimes advantageous to use graphical methods. Graphical methods are particularly useful when the data being operated on can not easily be expressed mathematically. Common examples are problems concerning irregular areas such as natural lakes and land masses. Other problems include empirical relations between temperature, time, force, displacement, velocity, and acceleration. Another group of problems involve social and economic studies such as population statistics, spread of disease, and immunity. In all these cases the data may be quite non-uniform and yet we wish to determine rates of change and the total or integrated effect. When the standard mathematical methods become too time consuming or involved, the graphical methods can produce results quickly and usually with an acceptable accuracy.

16.1 GRAPHICAL DIFFERENTIATION

Differentiation may be considered to be the determination of the rate of change of one variable with respect to a related variable. In calculus the slope at any point on a curve of $y = f(x)$ is expressed as

$$\frac{dy}{dx} = \lim_{\Delta x \to 0} \frac{\Delta y}{\Delta x}$$

where $\frac{dy}{dx}$ is called the derivative of y with respect to x. This is illustrated in Figure 16.1 for the slope at point A on the curve. The slope is represented by the tangent to the curve at this point.

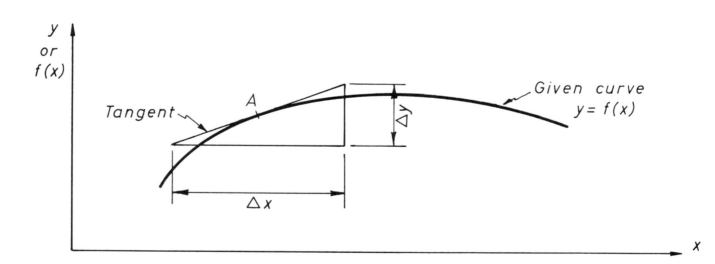

Figure 16.1 Slope of a Curve at a Point

In most applications not only is the slope required at a point on the curve, but over the entire useful range of the curve. What is required is a new function, the derived curve, showing the slope at all points along the curve. This derived curve is usually plotted below the given curve as shown in Figure 16.2. In this case the slope of the given curve is found at a number of points and then each point is plotted below in order to produce the derived curve. It should be noted that the vertical (ordinate) scale of f(x) and the horizontal (abscissa) scale of x may be different. In addition the scale of $\frac{dy}{dx}$ will almost certainly be different from the scale of x.

It can be seen that the process of differentiation includes finding the slope at a number of points on a curve. Methods of finding the slope will be considered next.

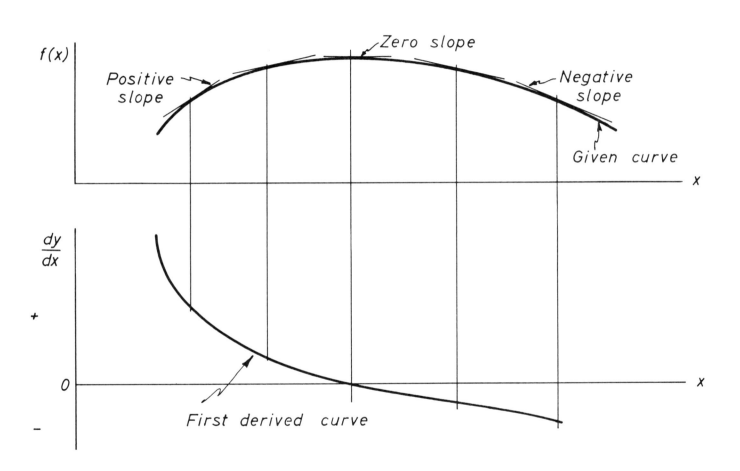

Figure 16.2 Given Curve and First Derived Curve

16.2 DETERMINATION OF SLOPE

It is difficult to determine the slope of a curve accurately by graphical methods. The result is that the graphically derived curve has a low degree of accuracy. Therefore it is desirable to have more than one method of finding the slope so that a check can be made on the results if necessary. We will consider two different methods of finding the slope (the tangent to a curve at a point).

16.3 REFLECTING SURFACE METHOD

In this method the normal to a curve is obtained by using a small reflecting surface such as an erasing shield or a mirror. The reflecting surface is held perpendicular to the paper at the point on the curve where the tangent is required, see Figure 16.3. The mirror or shield is then rotated about the point until the visible portion of the curve is reflected back on itself. The correct position occurs when the reflection and the visible curve form a smooth continuous line. The reflecting surface must be held perpendicular to the paper. When the curve and its reflection form a smooth continuous line then the reflecting surface is normal to the curve at this point. The required tangent is, of course, at right angles to the normal.

The reflecting surface can be used to obtain the normal to a curve and then a drafting triangle can be used to draw the tangent. However it is not necessary to actually draw the tangents at this stage, as will be shown later. A series of normal lines can be drawn at points along the curve as shown in Figure 16.4. The method is reliable and quite accurate.

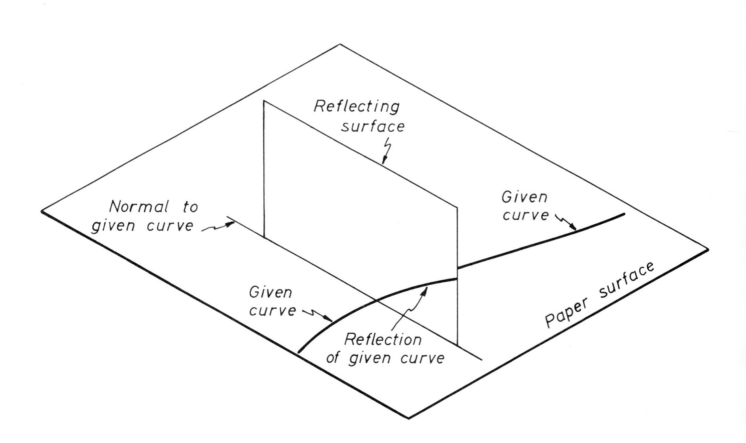

Figure 16.3 Reflecting Surface for Slope Determination

A decision must be made as to where to draw these normal lines along the length of the given curve. The positions are chosen using the rule that the points are chosen closer together in regions where the slope changes rapidly. Enough points must be selected to enable a smooth curve to be drawn for the derived curve.

A method of producing the first derived curve from the series of normal lines will be described. For convenience the derived curve will be plotted on top of the given curve. However it is usual to plot the derived curve below the given curve.

The given curve is shown in Figure 16.5 with the normals drawn at selected points.

Lay out a point P at a distance d to the left of the origin. For purposes of illustration let this distance be equal to one unit of the horizontal scale. The distance d is called the pole distance.

From P draw lines perpendicular to the normal lines as shown. These lines are parallel to tangents to the curve. Consider point M. The slope of PM is equal to the slope of the curve at N. Since OP = 1 unit, and the slope = $\frac{OM}{OP}$ then the distance OM is equal to the slope of the curve.

The scale for the $\frac{dy}{dx}$ ordinates can therefore be calibrated to read the slope directly.

The ordinate representing the slope (for example OM) at each point on the curve is next projected horizontally to its appropriate position for that point. This will result in a series of points which may be joined by a smooth curve to produce the required derived curve.

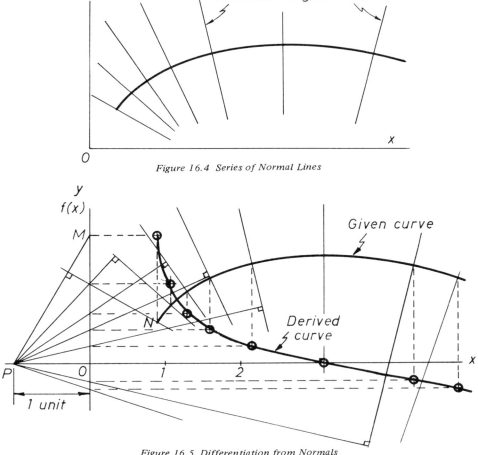

Figure 16.4 Series of Normal Lines

Figure 16.5 Differentiation from Normals

The choice of the pole distance equal to one unit of the horizontal axis may result in a derived curve which is too large or too small. Any convenient distance can be chosen, moving the pole farther out will increase the height of the curve and vice versa. The magnitude scale of the vertical axis is merely increased or decreased in the same ratio that the pole distance was increased or decreased. This process is illustrated in Figure 16.6 in which the pole distance was doubled.

16.4 CHORDAL METHOD

In this method the slope of the chord between two points close together on the curve is used to approximate the slope of the curve, Figure 16.7. The two points must be close enough together so that the portion of the curve between them is approximately an arc of a circle. The chord is bisected to locate the point on the curve where the tangent touches.

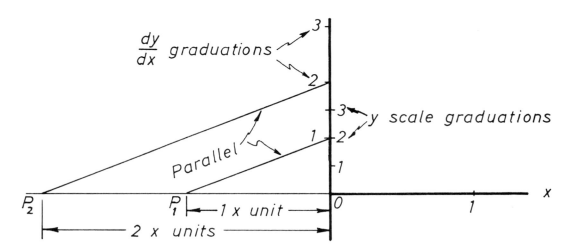

Figure 16.6 Effect of Changing the Pole Distance

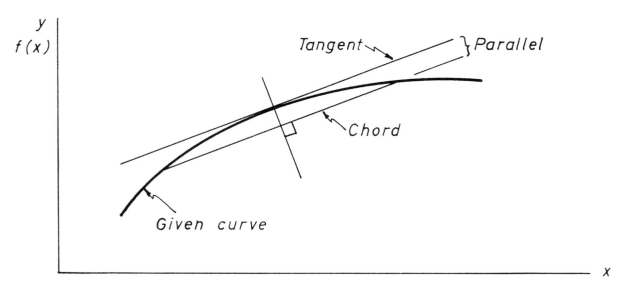

Figure 16.7 Chordal Method

The complete procedure is detailed below and illustrated in Figure 16.8.

Divide the given curve into a series of vertical strips. Make the strips narrow where the slope changes rapidly.

Draw tangents to the curve between each segment. The mirror method or the chordal method can be used. To avoid confusion letter the points of tangency A, B, C, etc.

Lay off a convenient pole distance $OP = d$ to locate the pole point P. A "convenient" distance is one which will result in the derived curve just filling the vertical space available and having a rational ordinate scale. The vertical height depends directly on the maximum slope and the pole distance d.

Through P draw rays parallel to the tangents. Let these rays intersect the vertical axis at the points a_1, b_1, c_1, etc. corresponding to the tangents at A, B, C, etc.

Locate the required point on the derived curve by projecting horizontally from a_1 to intersect a vertical line through A at the point A'.

For the calculation of the scale of the derived curve and for proof of the validity of the construction consider similar triangles $0Pa_1$ and MAN:

$$\frac{dy}{dx} = \frac{\Delta y_A}{\Delta x_A} = \frac{MN}{AM} = \frac{0a_1}{P0} = \frac{y_1}{d}.$$

Thus the vertical scale of the derived curve is equal to the scale of y divided by the pole distance d in units of the horizontal scale.

Locate the remaining points on the derived curve by repeating the construction above for each tangent drawn.

Draw a smooth curve through the derived points to produce the derivative curve.

In differentiation an ordinate of the derived curve represents the slope of the given curve at that point.

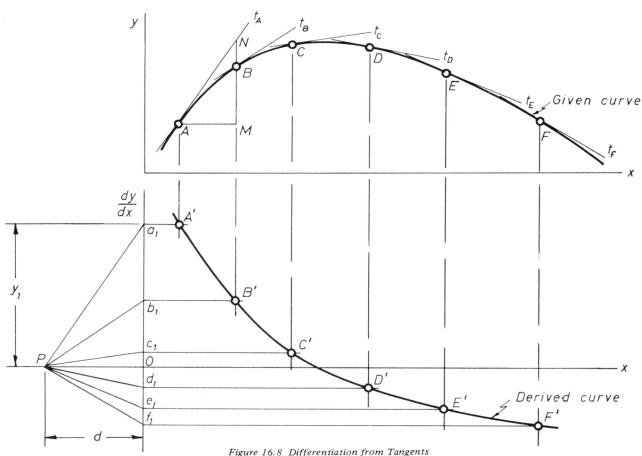

Figure 16.8 Differentiation from Tangents

16.5 GRAPHICAL INTEGRATION

From elementary calculus, an integral represents the area bounded by the curve being integrated, the x axis, and two ordinates. The process of integration is a procedure for summing many small areas of the above type. Graphical integration can be much more accurate than graphical differentiation. This is because the determination of an area can be done more accurately than the determination of the slope of a curve.

Integration may be considered to be the reverse of differentiation as we wish to find the curve from which a given curve was derived. That is, the ordinate at a point on the given curve represents the slope at that point on the required curve. For example in Figure 16.9 the ordinate at the point A' on the given curve represents the slope at that point on the integral curve. It can be seen that an infinite number of curves could have the slope as indicated on the graph above the given curve. The correct curve to select from this family of possible curves is determined by the boundary conditions of the particular problem. When using a purely mathematical technique for integration, the selection of the correct curve is made by using a constant of integration.

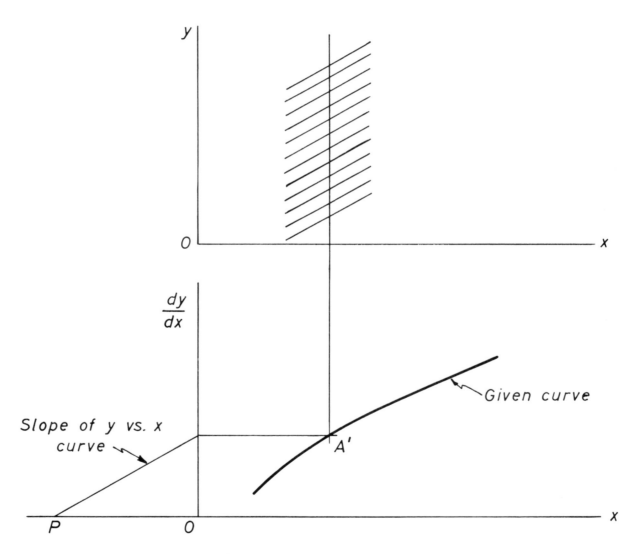

Figure 16.9 Slope and Boundary Conditions

16.6 MEAN ORDINATE–CHORDAL METHOD

The curve to be integrated is divided into a number of sections as shown in Figure 16.10. The area of a section under the curve is approximated by the area of the rectangle ABCD. The area ABCD will be exactly equal to the area under the curve if the two small areas AEF and FGB are equal.

In graphical integration these two small areas are estimated by eye. This can be done accurately by eye if the curve does not change shape too much in the interval, Δx. The eye is an accurate instrument for detecting differences when the shapes are similar. For this reason the vertical strips into which the curve is divided should be close together in regions where the curve changes shape rapidly.

The complete integration procedure includes the determination of the area of each section and adding these all together. This can be done by the following graphical procedure illustrated in Figure 16.11.

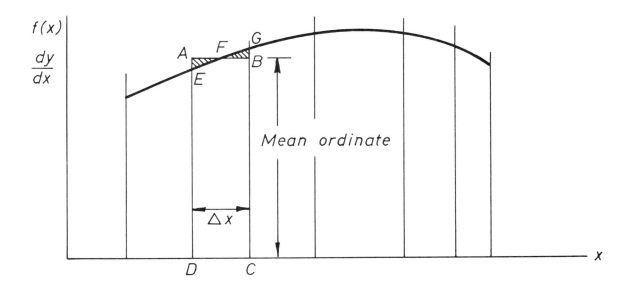

Figure 16.10 Area under Section of Curve

Divide the area under the given curve into a series of vertical strips.

By eye determine the mean ordinate for each strip. Project these ordinates to meet the vertical axis. To avoid confusion number these intersections 1, 2, 3, etc.

Lay off a convenient pole distance OP = d. A "convenient" distance will result in a graph filling the available space and having a rational scale. The maximum height will be equal to the total area under the given curve. Estimate this area and then choose d.

Join P to each of the intersections 1, 2, 3, etc.

To determine the first point on the integral curve draw a line OA through the origin parallel to ray P1. This assumes that the boundary condition for the problem includes zero at the origin of the integral curve. This line intersects the vertical boundary of the strip at point A.

To determine the scale of the integral curve and for proof of the validity of this construction consider similar triangles OP1 and OVA:

$$\frac{y_m}{y_1} = \frac{PO}{OV} = \frac{d}{k}.$$

The area of the strip = $y_m \times k = y_1 \times d$.
Thus the vertical scale of the integral curve is equal to the scale of y multiplied by the pole distance d in horizontal units. It can be seen that a larger value of d would decrease the height of the new curve, and vice versa.

The area of the next segment can be determined in the same manner as above. Through point V draw VW parallel to P2. The area of this strip is $y_2 \times d$. However this area must be added to the previous area. Therefore through A draw AB parallel to VW. Then the sum of the areas of the first two strips is $(y_1 + y_2) \times d = BR$. Therefore it is not necessary to draw VW since AB can be drawn directly to find point B.

Proceed in a similar manner to add all the areas together to find a succession of points on the curve. These points are considered to be the ends of chords.

Draw a smooth curve through the points A, B, etc. to form the integral curve.

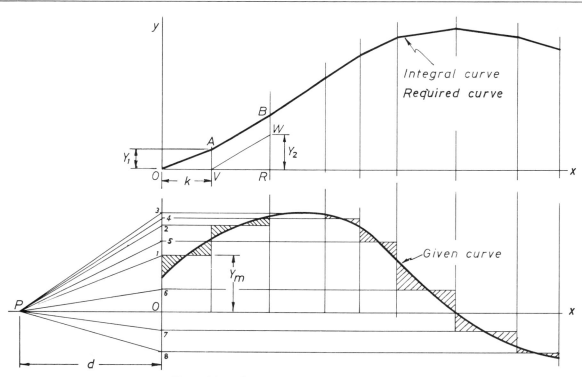

Figure 16.11 Graphical Integration of a Given Curve

Example 16.1 *Graphical Integration and Differentiation*

The data below was obtained from a test run of a vehicle and provides a record of the velocity in metres per second (m/s) as a function of the time after launch in seconds.

Time	0	20	40	60	80	100	120	140	160	180	200	220	240	260
Velocity	0	3	10	20	32	42	47	47	43	35	24	12	4	0

a) Plot the data and draw a smooth curve through the points. Use scales of 1 mm = 1 m/s and 1 mm = 2s. This is the given curve.
b) Graphically integrate the given curve to produce a curve of displacement vs time.
c) Graphically differentiate the given curve to generate a curve of acceleration.

Solution The solution below has been reduced in size from its original size.

a) Plot the velocity-time graph in the middle of the working area because the displacement curve will be plotted above and the acceleration curve below the velocity curve.
b) Estimate the area under the given curve to be (25)(260) = 6500 m. To fit this in the available space of 65 mm a scale of 1 mm = 100 m is required. Therefore choose a pole distance of 50 mm (50 x horizontal scale of 2 = 100). Divide the given curve into strips and use the mean ordinate chordal method as shown.
c) The maximum slope is estimated to be about 0.5 m/sec². Available space is approximately 25 mm. A suitable scale is 1 mm = 0.02 m/sec². Therefore the pole distance is 1/0.02 = 50 mm in horizontal units. Therefore choose the pole distance of 25 mm. Draw the slope at selected points. Project to develop points on the curve. Join the points with a smooth curve.

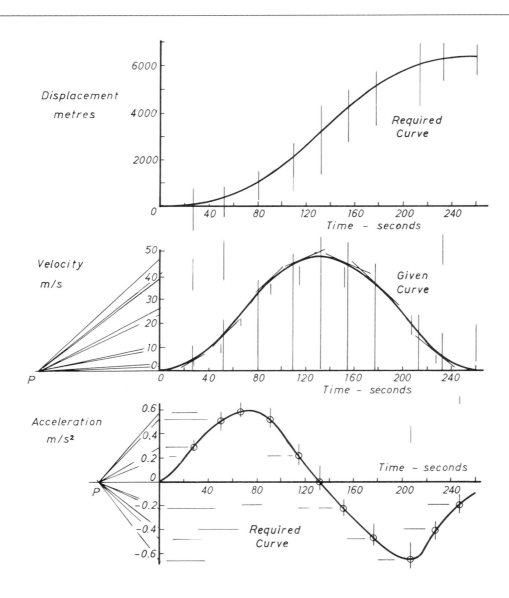

PROBLEMS

P 16.1 The data shown below was obtained from a test run of a vehicle and show the velocity after starting the run at various times. Plot the data and graphically differentiate it to obtain the acceleration vs time graph. Use a velocity scale of 1 mm = 0.1 m/sec and a time scale of 1 mm = 1 sec. Choose a suitable scale for the acceleration ordinates. What is the value of the maximum acceleration, minimum acceleration and acceleration after 40 seconds?

P 16.2 Integrate the data of problem 16.1 to find the distance travelled as a function of time. Choose a suitable scale for the distance ordinates. What is the distance travelled after 65 secs? What is the length of the runway?

Time Seconds	Velocity Metres/sec.
0	0
15	0.85
30	2.95
45	5.6
60	6.35
75	5.1
90	3.1
105	1.3
120	0.3
135	0

P 16.3 The population data for a town is given below as a function of time in years. Plot the data and draw a smooth curve to connect the points. In what year did the most rapid growth occur? Choose suitable scales.

Year	Population
1900	1500
1905	1300
1910	1300
1915	1700
1920	2300
1925	3250
1930	3900
1935	4300
1940	4600
1945	4700
1950	4650
1955	4350
1960	4000
1965	3550
1970	3250

P 16.4 A force acts in the same direction as an object which moves along a straight line. The variation of force and distance moved is given by the data shown below. Draw a curve to show the relation between the work done in Newton metres and the distance moved. Choose suitable scales.

Force-Newtons	24.0	19.7	16.2	11.5	8.6	7.0	5.7	5.0	4.3
Distance-Metres	0	0.15	0.30	0.60	0.90	1.2	1.5	1.8	2.1

P 16.5 The vertical displacement of the cam follower, Figure P 16.5, for one revolution of the cam is given in the table below. The cam makes one revolution every 2.0 seconds. Plot the data showing displacement against time. Differentiate once to produce a velocity vs time graph.

θ degrees	Displacement mm	θ degrees	Displacement mm
0	0	195	36.2
15	2.5	210	32.5
30	6.3	225	26.8
45	12.0	240	18.7
60	18.7	255	12.0
75	26.8	270	6.3
90	32.5	285	2.5
105	36.2	300	0.0
120	37.5	315	0.0
135	37.5	330	0.0
150	37.5	345	0.0
165	37.5	360	0.0
180	37.5		

Figure P16.5

P 16.6 A plot of land borders on a stream as shown in Figure P 16.6. Determine the area of the land. Draw property lines, perpendicular to the base line, which will divide the plot into three equal area lots.

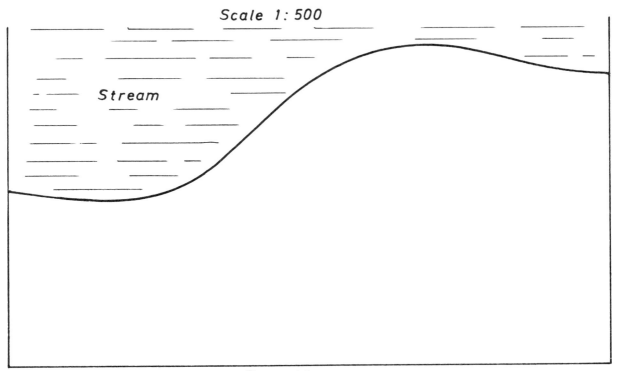

Figure P16.6

17. Short-Range Photogrammetry

Photogrammetry may be defined as the science of the transformation of photographs into maps. It is a well established and highly technical subject, the photographs normally being aerial views taken from a considerable altitude. In this section we will consider the transformation of close-range photographs into orthographic drawings.

Short-range photogrammetry is useful in engineering when direct measurement is impossible or too time-consuming for practical purposes. The photograph can be measured to provide the information which otherwise would require more time and effort. In some cases a photograph may be the only practical way to record a physical situation.

17.1 THE PHOTOGRAPHIC VIEW

A camera will record on the film a view which is a perspective view. Referring to Section 12.9, Perspective Projection, the camera lens can be considered to be located at the Station Point. The resulting photograph would be similar to the view we obtain on the vertical picture plane if the film in the camera is vertical. It should be noted that the camera may be tilted and then the film will not be vertical. This case will be considered later.

The photograph will be of limited usefulness unless we have some means of scaling the photographic print obtained. Therefore, we provide some sort of scale on the object. The photograph, showing the scale, can be measured and is called a "controlled" photograph.

The photograph is a perspective view and we wish to change this into an orthographic view. That is, we reverse the procedure used in producing a perspective view. In Chapter 12 of the text we started with orthographic views and produced a perspective view.

The particular situation illustrated in Figure 17.1 shows a case where the camera is held in a horizontal position, that is, the film plane is vertical. In addition in this case the object being photographed is a vertical wall. The wall has a series of squares marked on it which are reproduced on the photograph as an array of trapezoids. It can be seen that in the vertical direction, equal divisions on the wall will become another set of equal divisions on the photograph. This particular condition holds true for this case of a vertical object and vertical film in the camera. This fact can be used to facilitate the graphical construction to be described in the following example.

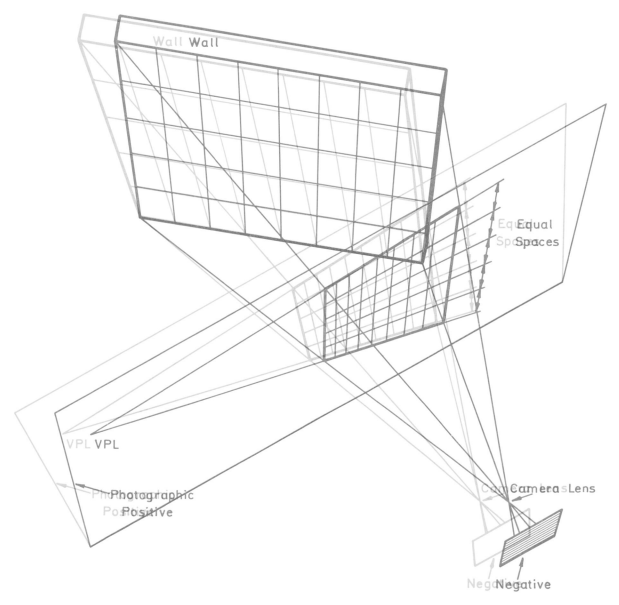

Figure 17.1 Horizontal Photography of a Vertical Object

Example 17.1
Consider the photograph shown in Figure 17.2. The wall in the background is obstructed because of the presence of the research equipment in front of it. Note that there is a set of axes in the form of a square 1.5 m-by-1.5 m placed on the wall. This is a controlled photograph.

The problem posed is that it is required to run a pipe from point X up to Y at the top of the conduit and then over to exit at a point, Z, mid-way between the two ventilators V and W. We want to know the length of the pipe and the angle of the bend at Y.

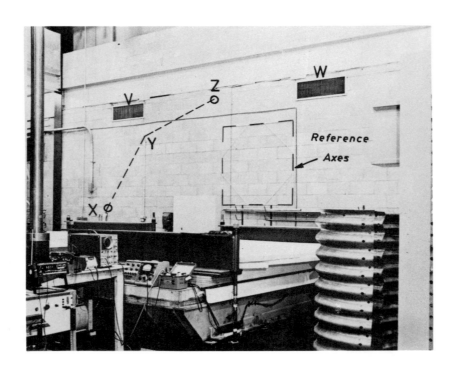

Figure 17.2 Photograph showing Wall-Piping Requirements

From your knowledge of perspective projection you may recall that any set of parallel lines on the object will meet at a vanishing point. (Note that these lines do not necessarily meet at the vanishing points VPR and VPL which were established in Section 12.12). We can draw lines through the parallel sides of the 1.5 m square as shown in Figure 17.3. The image of the square itself becomes a trapezoid on the print.

Note that in this photograph the vertical lines are parallel. This indicates that the vertical vanishing point is at infinity upwards and downwards.

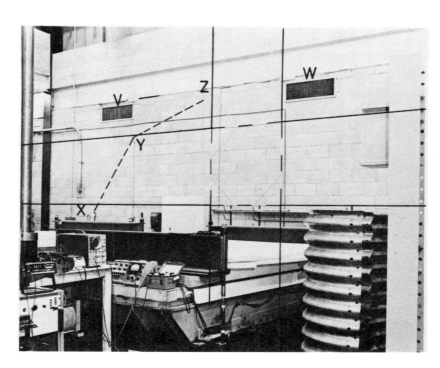

Figure 17.3 First Trapezoid Extended

Making use of the principle of similarity, we can construct a series of trapezoids as shown in Figure 17.4. Each trapezoid represents a 1.5 m square on the wall.

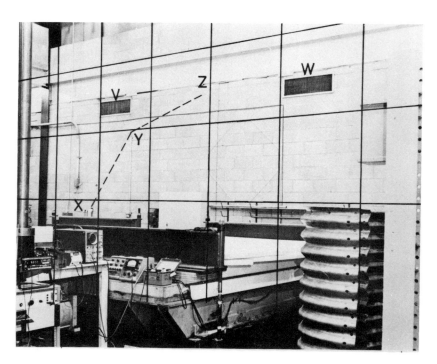

Figure 17.4 Trapezoids Forming a Grid on the Wall

The method of constructing the grid of trapezoids will be explained by reference to Figure 17.5. The explanation is for a general situation in which the camera back and film are not parallel to the object plane.

Extend the sides of the reference square DA and CB to meet at a vanishing point F. All lines on the photograph parallel to DA and CB will also meet at F. Extend the sides BA and CD to meet at the vanishing point E. All lines parallel to BA and CD will also meet at E.

We will now construct a series of trapezoids, each one representing a 1.5 m square. Let the diagonals DB and CA intersect at G. Through G draw extended lines to E and F. Mark intersection points 1, 2, 3, and 4.

Extend line C–1 to H, D–1 to J, D–2 to K, and A–2 to L. This construction forms two more trapezoids ABJH and CBKL each representing a 1.5 m square.

The procedure can be repeated or the diagonals of trapezoids can be extended to produce a grid of 1.5 m squares covering the complete area required as shown in Figure 17.4.

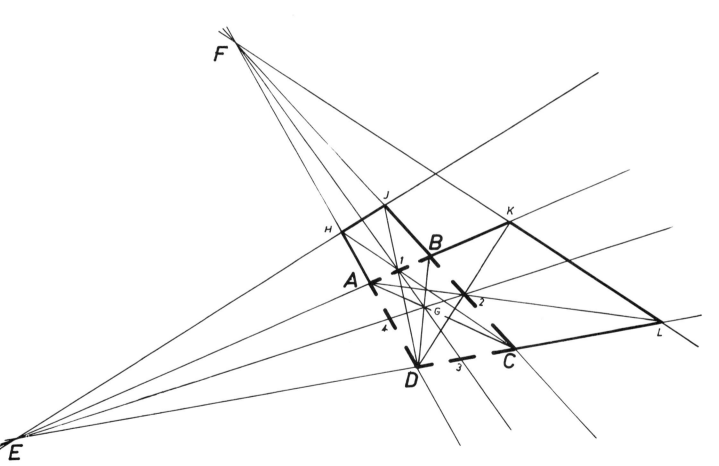

Figure 17.5 Method of Extending Trapezoids

Note that each side of the original 1.5 m square of Fig. 17.2 had been divided into 5 equal divisions each 0.3 m long. These 0.3 m divisions on the first trapezoid can now be used to further subdivide the photograph. Draw lines to the two vanishing points through each 0.3 m mark. This will produce a 0.3 m grid inside the first trapezoid. The 0.3 m grid can be expanded beyond the first trapezoid by using the intersection of these 0.3 m lines and the diagonals of the large trapezoids previously drawn. The final grid is shown in Figure 17.6. This is known as "gridding" the photograph. As a check on the graphical work remember that in a horizontal photograph of a vertical object, equal vertical divisions on the object will result in equal vertical divisions on the photograph.

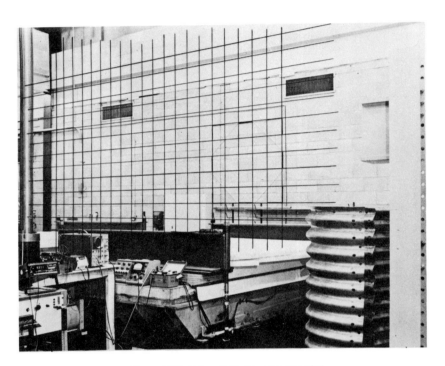

Figure 17.6 0.3 m Grid Covering the Wall

Now that the photograph is covered by a grid representing 0.3 m squares, an orthographic view can be made. This drawing is shown in Figure 17.7 and was produced from an 8" x 10" photograph. The orthographic drawing can be scaled to determine the dimensions required. In this example xy = 1.88 m, yz = 1.55 m and the angle xyz = 149°.

The accuracy obtained depends on the size of the working photograph and the care exercised in the graphical constructions. One method of drawing the trapezoids is to tape a sheet of thin acetate over the photograph and scribe the required lines using a sharp steel scriber.

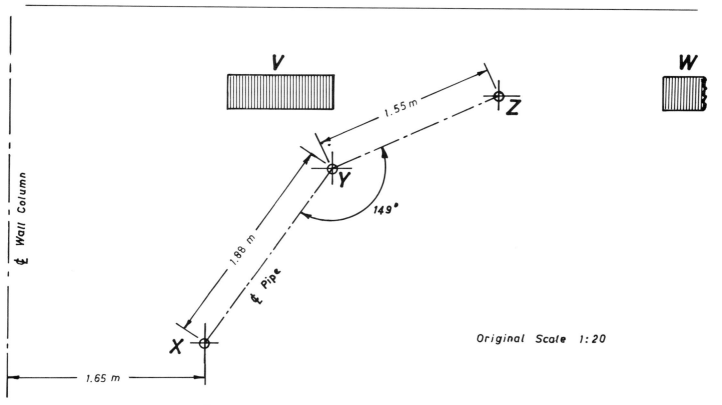

Figure 17.7 Orthographic Drawing of Wall from Scaled Photograph

PROBLEM P 17.1

The photograph on the next page shows a portion of the wall in a laboratory. We want to run a pipe from the point R up to the point S and then over to an outlet at point T.

The axes shown in the photograph are actually 1.5 m square and have 0.3 m divisions indicated.

Using the methods described for short range photogrammetry produce an orthographic view of the wall showing the significant features. Find the lengths RS and ST of the pipe required and the angle of the bend at S.

Procedure

1. Tape the photograph to the extreme left hand side of your drawing board as the vanishing point will be about 450 mm to the right of the right hand side of the photograph. It is convenient to orient the photograph so that the upright sides of the reference axes are vertical. The textbook may be folded back to permit the photograph to lie flat on your drawing board.

2. Tape a sheet of tracing paper over the photograph.

3. Draw the trapezoid representing the data axes and extend to find the vanishing points.

4. Extend diagonals to produce surrounding trapezoids representing 1.5 m squares.

5. Use the 0.3 m divisions to further subdivide the surface of interest.

6. From this final grid, draw an orthographic view of the wall area required. Measure the required lengths and angle. The orthographic drawing should be to a scale of 1:20.

Science and Technology Division
CAMOSUN COLLEGE
1950 LANSDOWNE ROAD
VICTORIA, B.C.   V8P 5J2

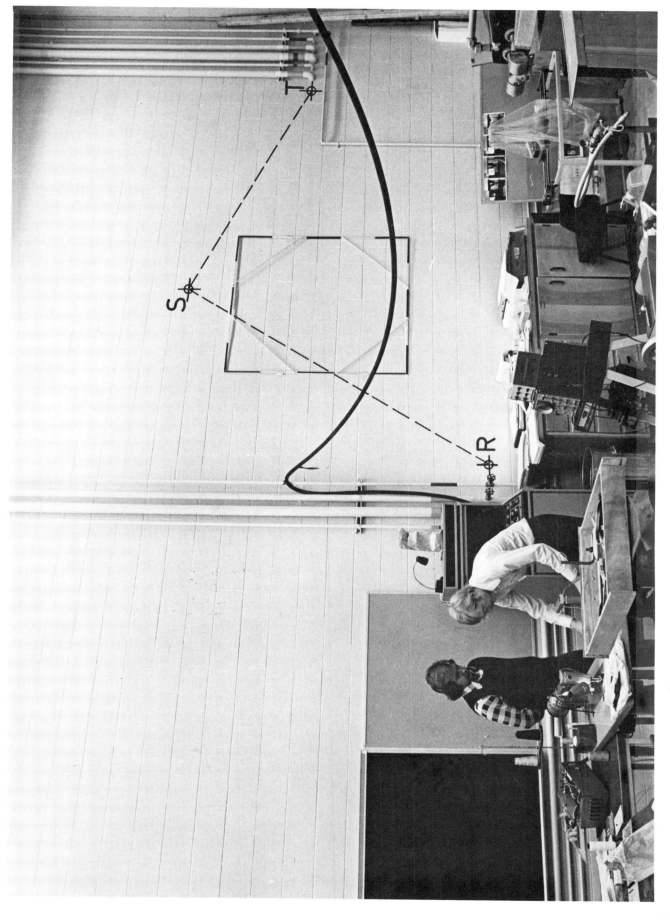

Problem P17.1

17.2 OBLIQUE PHOTOGRAPHY OF FLAT GROUND

Another use of short-range photogrammetry involves the photography of flat ground from a low flying aircraft or from nearby high ground. In this case the camera is at an angle and the resulting photograph is called a low oblique. The situation is shown in Figure 17.8 in which the ground has been marked out in regular squares. The resulting photograph will show an array of trapezoids. Note that in this case that along any line parallel to the horizon on the photograph equal spaces on the ground will be represented by equal spaces between the sides of the trapezoids. This fact is useful in controlling the graphical constructions.

The horizon may not show on the photograph but two points on it, VP1 and VP2 may be found by extending the sides of one datum trapezoid as shown before in Figure 17.5. In Figure 17.5 the horizon would be a line joining points E and F. As in the previous section a datum set of axes is required on the object. In this case it can be an accurate square laid out on the ground prior to taking the photograph. The procedure is illustrated in the following example.

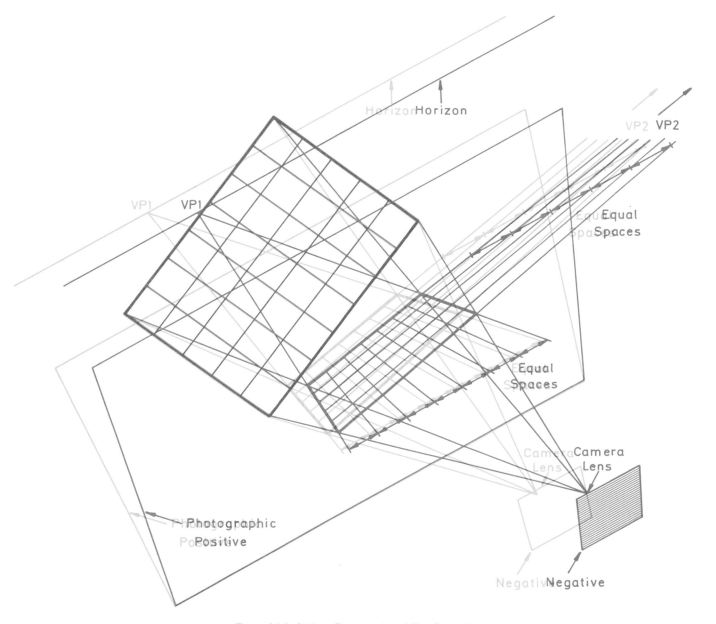

Figure 17.8 Oblique Photography of Flat Ground

Example 17.2
The photograph, Figure 17.9, was taken showing a simulated archeological site as it might appear from an aircraft at a height of about 60 metres. The site consists of a series of ancient stone monuments arranged in some sort of pattern on flat ground. In addition there are some remains of buildings to the right of the erect stones. A reference square has been laid out on the ground using 0.3 m-square white plywood markers. The four white markers have been accurately spaced to form the sides of a large square 6 m by 6 m.

You are required to make an orthographic plan of the simulated archeological site using the methods developed in this chapter.

Solution
Tape the photograph on to the drawing board and tape a sheet of drawing paper over it.

Draw fine lines through the centres of the white dots and extend these lines to produce the first trapezoid. The two vanishing points will also be found where parallel sides of the square meet. These vanishing points will be above and to the left and right of the photograph. Draw in the horizon line.

Extend the first trapezoid to form the array of trapezoids covering the area of interest on the photograph. Use the method illustrated in Figure 17.5. As a control and check on your array use the fact that along any line parallel to the horizon equal spaces should be marked off by the trapezoidal lines produced to the vanishing points.

The final graphical work is shown in Figure 17.10.

From the array of trapezoids representing 6 m squares on the ground, draw an orthographic plan view. This orthographic drawing is shown in Figure 17.11 drawn to a scale of 1:240. It can be seen that the stone monuments were laid out in the form of an ellipse.

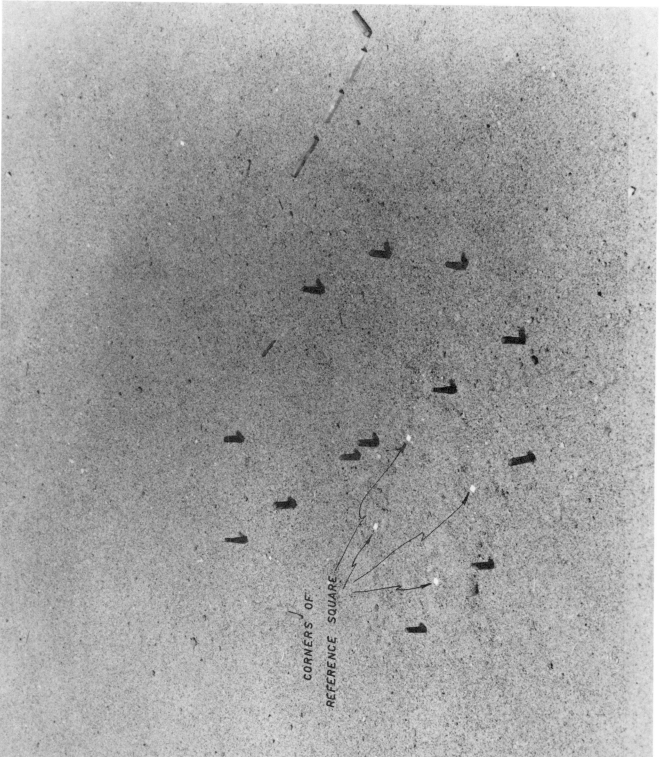

Figure 17.9 Oblique Photograph of Simulated Site

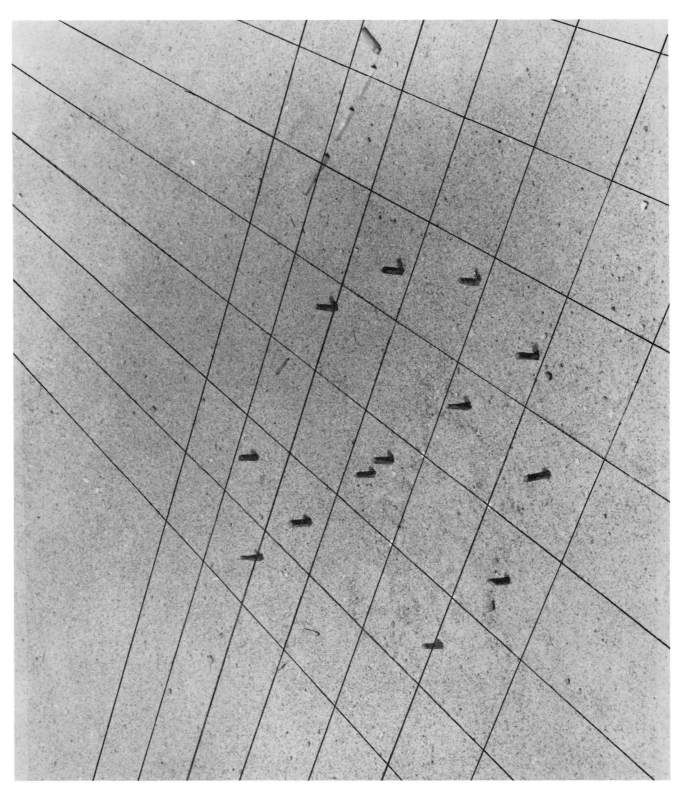

Figure 17.10 Gridded Photograph

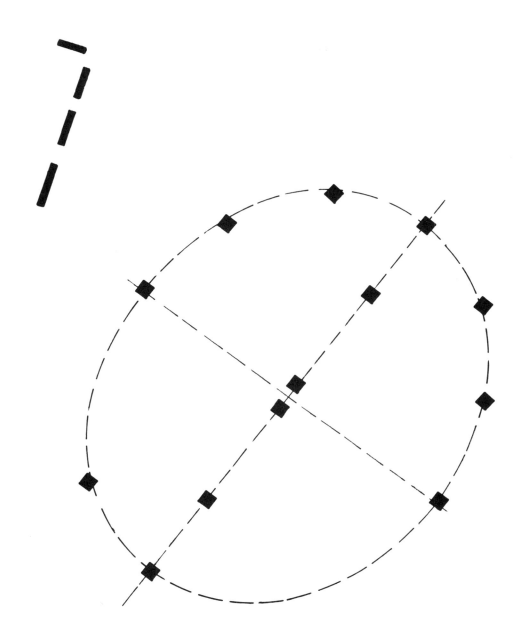

SCALE 1:240

Figure 17.11 Orthographic View of Simulated Archaeological Site

PROBLEM P 17.2

The drawing below represents a photograph of a large wall along which it is proposed to run a pipe from A to B to C to D. The reference frame shown is 5 metres square with one-metre divisions. Produce an orthographic view of the wall showing significant features. Determine the lengths AB, BC, and CD and the angles of the bends at B and C.

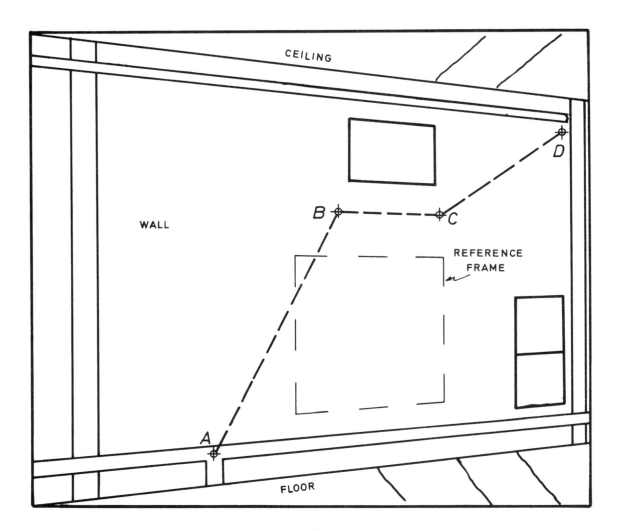

Figure P17.2

P 17.3 A simulated archeological site is shown in Figure P 17.3 as it might be photographed from a low-flying aircraft. A reference square 10 metres on each side has been laid out on the ground, the corners being identified by the centre of the cross within the triangles △. The plain square marks □ represent ancient stone monuments. Prepare an orthographic plan of the layout of the stones.

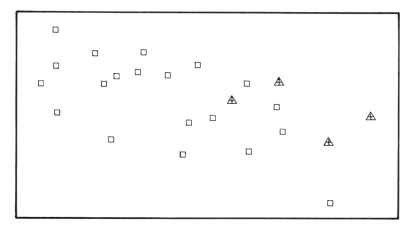

Figure P17.3

P 17.4 An aerial photograph of part of a coastline is simulated in Figure P 17.4. The reference square is 20 metres △. Prepare a reasonable orthographic plan of the shoreline. What is the distance from the end of the pier P to the lighthouse L?

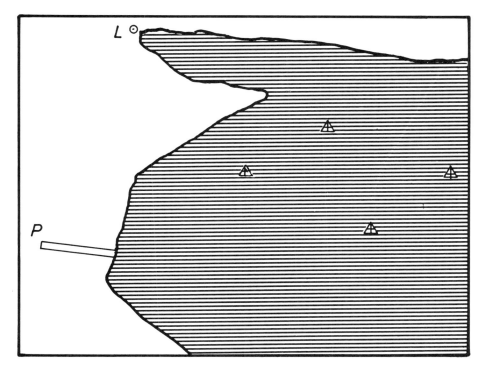

Figure P17.4

18. Computers, Calculators, and Graphics

COMPUTERS, CALCULATORS, AND PLOTTING MACHINES

Simple hand calculators can be used to facilitate the production of some types of pictorial views. More elaborate programmable calculators and plotting machines will reduce the time and effort required to an even greater degree. A procedure will be developed for producing isometric and perspective views using simple techniques and equipment.

18.1 GENERAL PROCEDURE

The normal orthographic horizontal and frontal projections are the starting point. These two views provide the location of any point on or within the object to be portrayed. The first step is to choose a reference datum and express each significant point on the object in terms of X, Y, and Z coordinates. This data is then operated on using a hand calculator, programmable calculator, or a computer. The results are plotted, either by hand or using a plotting machine. The significant points are then joined by lines to form the desired isometric view. The detailed procedure is outlined below for normal isometric views and for perspective views.

18.2 NORMAL ISOMETRIC VIEW

In the normal isometric view the isometric plotting axes can be arranged as shown in Figure 18.1. The location of any point $X_1 Y_1 Z_1$ is obtained by measuring along the axes as shown to arrive at the final position P.

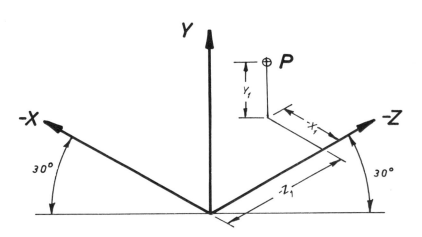

Figure 18.1 Normal Isometric Axes

Plotting machines generally use two axes, horizontal, x, and vertical, y. Also for hand plotting on a drafting board it is convenient to use these axes. Therefore we wish to plot using the x and y axes as shown in Figure 18.2. The position of the point P can be determined from trigonometry (as shown in Figure 18.2).

From Figure 18.2 the x and y coordinates of point P are:

$$x = -Z_1 \cos 30 - (-X_1 \cos 30)$$
$$= 0.866 (X_1 - Z_1)$$
$$y = -Z_1 \sin 30 - X_1 \sin 30 + Y_1$$
$$= -0.5 (X_1 + Z_1) + Y_1$$

If it is desired to make an isometric projection of an object the expressions must each be multiplied by the foreshortening ratio, 0.816.

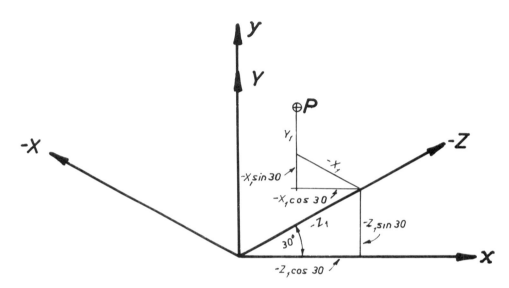

Figure 18.2 Plotting Axes and Isometric Axes

18.3 NORMAL ISOMETRIC VIEW-EXAMPLE

A normal isometric view will be generated for the block shown in orthographic projection in Figure 18.3.

Set up datum axes X_2, Y_2, and Z_2 at the convenient point 0_2 and record the coordinates taken from the orthographic views. These may be set up in a table to facilitate hand calculation using an electronic hand calculator as shown in Table 18.1. Enter the coordinates for each significant point, A,B,C, etc. on the block. Next perform the calculations filling in the columns as shown. Plot the x and y pairs of points and join the correct ones to produce the normal isometric view shown in Figure 18.4.

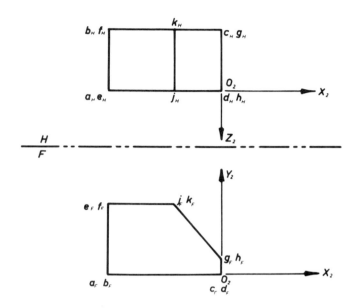

Figure 18.3 Orthographic Projections of Block

Point	Coordinates			$X_2 - Z_2$	$X_2 + Z_2$	$0.5(X_2+Z_2)$	x $0.866(X_2-Z_2)$	y $-0.5(X_2+Z_2)+Y_2$
	X_2	Y_2	Z_2					
A	-30.5	0	0	-30.5	-30.5	-15.25	-26.4	15.25
B	-30.5	0	-16.5	-14	-47	-23.5	-12.1	23.5
C	0	0	-16.5	16.5	-16.5	-8.25	14.3	8.3
D	0	0	0	0	0	0	0	0
E	-30.5	19.0	0	-30.5	-30.5	-15.25	-26.4	34.25
F	-30.5	19.0	-16.5	-14	-47	-23.5	-12.1	42.5
G	0	4.3	-16.5	16.5	-16.5	-8.25	14.3	12.6
H	0	4.3	0	0	0	0	0	4.3
J	-12.7	19.0	0	-12.7	-12.7	-6.35	-11.0	25.4
K	-12.7	19.0	-16.5	3.8	-29.2	-14.6	3.3	33.6

Table 18.1 Computation Table – Normal Isometric

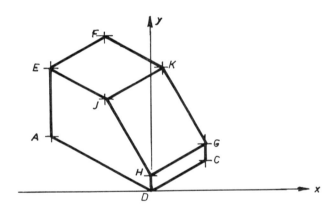

Figure 18.4 Normal Isometric Drawing

18.4 INVERTED ISOMETRIC VIEW

In some cases it is desirable to have a view of an object from below. In this case the isometric axes are arranged as shown in Figure 18.5.

In this case we could set up our datum axes X_2, Y_2, and Z_2 at the top of the block and obtain the co-ordinates for plotting. However, since we already have the data from our previous example it is convenient to use the data already available with the origin shown in Figure 18.3. The x coordinate for plotting will be the same as before.

$$x = 0.866(X_1 - Z_1)$$

The y coordinate however is measured up rather than down so that

$$y = 0.5(X_1 + Z_1) + Y_1$$

The computation table is shown in Table 18.2. The resulting inverted isometric drawing is shown in Figure 18.6.

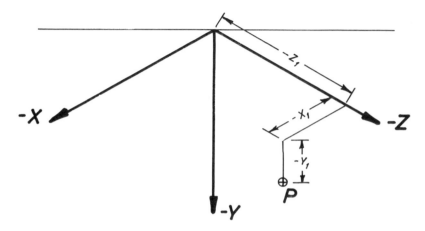

Figure 18.5 Inverted Isometric Axes

Point	Coordinates			$X_2 - Z_2$	$X_2 + Z_2$	$0.5(X_2+Z_2)$	x $0.866(X_2+Z_2)$	y $0.5(X_2+Z_2) + Y_2$
	X_2	Y_2	Z_2					
A	-30.5	0	0	-30.5	-30.5	-15.25	-26.4	-15.25
B	-30.5	0	-16.5	-14	-47	-23.5	-12.1	-23.5
C	0	0	-16.5	16.5	-16.5	-8.25	14.3	-8.3
D	0	0	0	0	0	0	0	0
E	-30.5	19.0	0	-30.5	-30.5	-15.25	-26.4	3.75
F	-30.5	19.0	-16.5	-14	-47	-23.5	-12.1	-4.5
G	0	4.3	-16.5	16.5	-16.5	-8.25	14.3	-4.0
H	0	4.3	0	0	0	0	0	4.3
J	-12.7	19.0	0	-12.7	-12.7	-6.35	-11.0	12.7
K	-12.7	19.0	-16.5	3.8	-29.2	-14.6	3.3	4.4

Table 18.2 Computation Table – Inverted Isometric

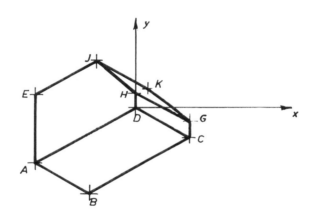

Figure 18.6 Inverted Isometric Drawing

18.5 COMPUTER PROGRAM-ISOMETRIC PROJECTION

For more complicated objects a large number of coordinate points are required. A programmable calculator or a digital computer will reduce the time required to calculate the x and y coordinates for plotting.

As an example isometric projections will be obtained for the object shown half full size in orthographic projection in Figure 18.7.

The computer program and print out is shown in Table 18.3. This is for a normal isometric projection so that the foreshortening ratio 0.816 has been used in the program. The x and y coordinates to be plotted are shown in the last two columns. The X_2, Y_2, and Z_2 data is printed in the first three columns to facilitate checking if an error appears to exist.

The resulting normal isometric projection is obtained by plotting the x and y pairs in the last two columns. The resulting isometric projection is shown in Figure 18.8 (half full size).

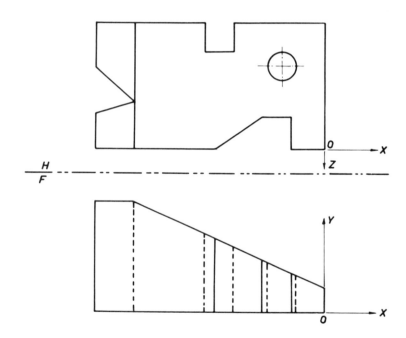

Figure 18.7 Figure in Orthographic Projection

```
        $JOB     WATFIV
1        5       READ,X,Y,Z
2                IF(X.EQ.400)STOP
3                A = 0.816*(0.866 *(H-Z))
4                B = 0.816*((-0.5 *(H+Z))+Y)
5        6       FORMAT(' ',5X,F7.1,5X,F7.1,5X,F7.1,5X,F7.1,5X,F7.1,5X,F7.1)
6                PRINT 6,X,Y,Z,A,B
7                GO TO 5
8                END

        $ENTRY
        -121.9      0.0        0.0      -86.2      49.7
        -121.9      0.0      -68.5      -37.7      77.7
        -121.9     61.0        0.0      -86.2      99.5
        -121.9     61.0      -68.5      -37.7     127.5
        -121.9     61.0      -19.3      -72.5     107.4
        -121.9     61.0      -44.2      -54.9     117.5
        -121.9      0.0      -19.3      -72.5      57.6

        -101.6     61.0      -25.4      -53.8     101.5
        -101.6      0.0      -25.4      -53.8      51.8
        -101.6     61.0        0.0      -71.8      91.2
        -101.6     61.0      -68.6      -23.3     119.2
        - 63.5     42.9      -68.6        3.6      88.9
        - 63.5      0.0      -68.6        3.6      53.9
        - 63.5     42.9      -53.1      - 7.4      82.6
        - 63.5      0.0      -53.1      - 7.4      47.6
        - 48.3     35.8      -68.6       14.4      76.9
        - 48.3      0.0      -68.6       14.4      47.7
        - 48.3     35.8      -53.1        3.4      70.6
        - 48.3      0.0      -53.1        3.4      41.4
```

| X_2 | Y_2 | Z_2 | x | y |

Table 18.3 Computer Program—Normal Isometric Projection

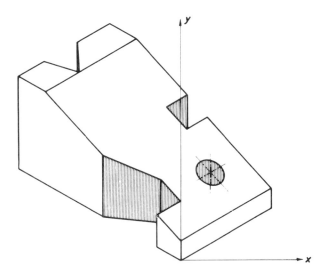

Figure 18.8 Normal Isometric Projection

An inverted isometric drawing of the same object is shown in Figure 18.9. The same orthographic position data as obtained for the normal isometric was used. The appropriate equations for an inverted isometric projection, as explained in section 18.4 were used in the computer program.

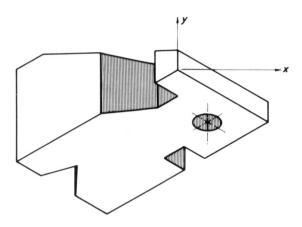

Figure 18.9 Inverted Isometric Projection

18.6 CHOICE OF AXES

It may be more convenient to choose axes so that all points on the object have positive coordinates. In addition it may be more familiar to designate the vertical direction as the Z axis. Such a system is illustrated in Figure 18.10.

The axes chosen previously, as in Figure 18.3 were selected so as to conform to the locating system developed earlier in the book (Section 4.2) for third-angle projection.

The mathematical expressions used in the production of an isometric view from orthographic data will be altered if the axes are changed. The basic method is the same as that previously described in Section 18.2.

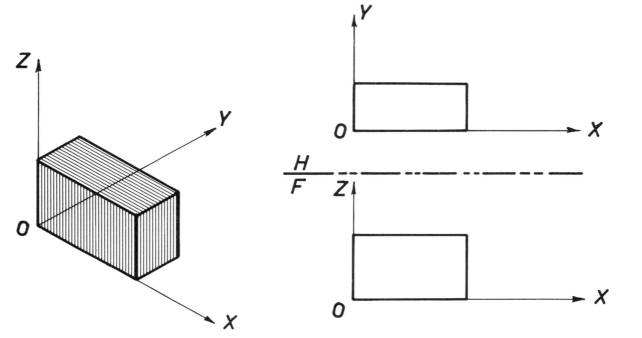

Figure 18.10 Alternate Choice of Axes

18.7 CLOSURE

The effort involved in the computation method of producing isometric views may be equal to the purely graphical method. However, once the position of each significant point on an object is obtained in XYZ coordinates, this data may be operated on to produce an infinite variety of views. For example the object can be tilted, rotated, moved backward or forward by a fairly simple mathematical transformation of the coordinates.

18.8 PERSPECTIVE PROJECTION BY COMPUTATION AND PLOTTING

The general arrangement of perspective projection was detailed in Chapter 12. The situation is shown in Figure 18.11 which is taken directly from Figure 12.18.

In a perspective view, each point on the object is projected on to the picture plane. Let the co-ordinates of a point on the object be X, Y, and Z. Also let the coordinates of this point on the picture plane be x and y. A convenient point for the origin of both these co-ordinates axes is the lower left-hand corner of the picture plane, 0.

For hand or machine plotting we require the co-ordinates x and y, for each significant point on the object whose co-ordinates are X, Y, and Z.

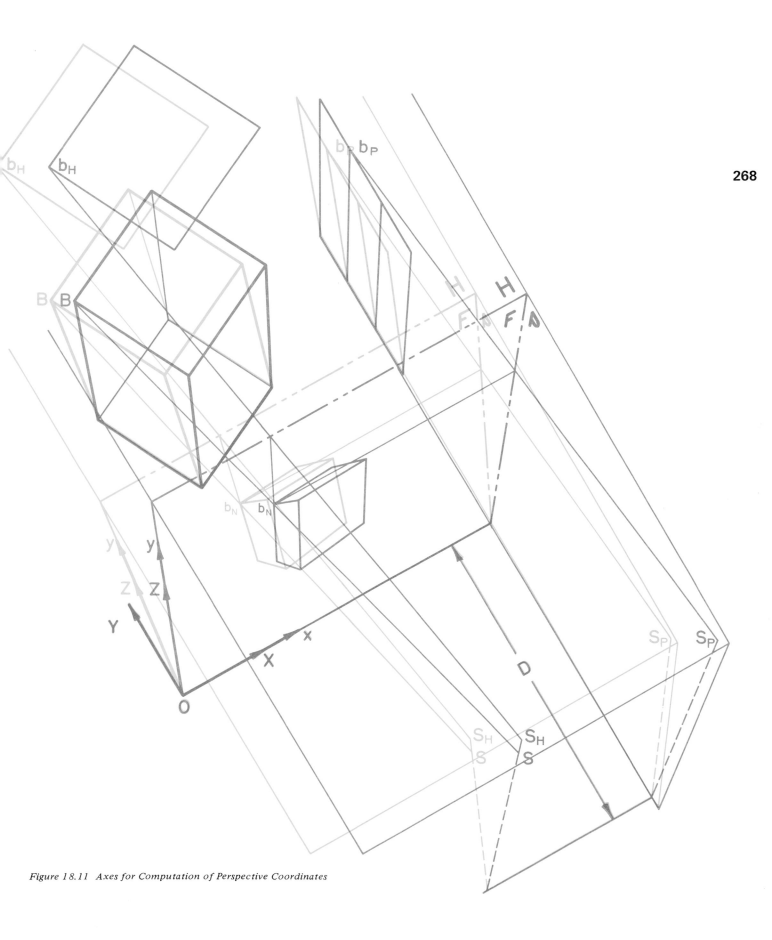

Figure 18.11 Axes for Computation of Perspective Coordinates

18.9 DETERMINATION OF PLOTTING COORDINATES

In perspective projection the location of the object, the picture plane and the point of sight must be decided before the drawing is made.

These variables, which affect the shape of the final drawing, are shown in orthographic projection in Figure 18.12. The location of plotting axes, x and y, and the object axes, X, Y, and Z are also shown.

Let D = distance of Station Point in front of the picture plane;
H = height of Station Point;
R = lateral distance of Station Point from origin.

Consider a general point B on the object. We wish to obtain the coordinates of the piercing point of SB on the picture plane. That is, we require x and y in terms of X, Y, and Z. This can be done by geometry as follows.

Consider similar triangles $b_H PQ$ and $b_H T s_H$

$$\frac{PQ}{Ts_H} = \frac{b_H P}{b_H T}$$

i.e., $\dfrac{x - X_1}{R - X_1} = \dfrac{Y_1}{D + Y_1}$

or $x = \dfrac{Y_1 (R - X_1)}{D + Y_1} + X_1$ \hfill (a)

From similar triangles $b_P MN$ and $b_P L s_P$

$$\frac{MN}{Ls_P} = \frac{b_P M}{b_P L}$$

i.e., $\dfrac{y - Z_1}{H - Z_1} = \dfrac{Y_1}{D + Y_1}$

or $y = \dfrac{Y_1 (H - Z_1)}{D + Y_1} + Z_1$ \hfill (b)

Now we have an expression for the plotting points in terms of the object points. Note that D is positive when S_H is placed below the axis 0X as in Figure 18.12.

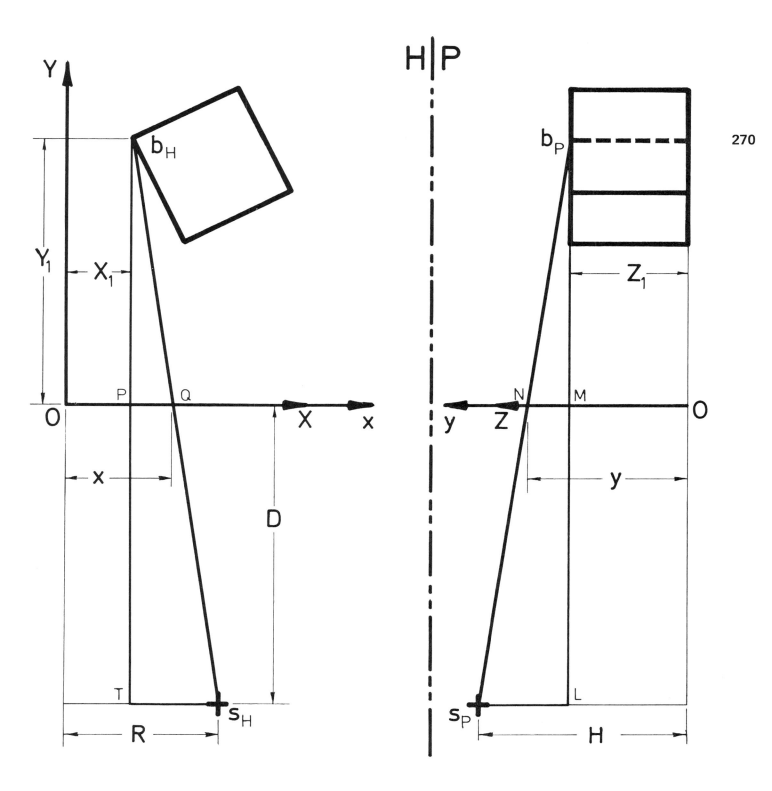

Figure 18.12 Geometrical Relations on Orthographic Views

18.10 PERSPECTIVE EXAMPLE

As an example we will make a perspective projection of the cube shown in Figure 18.11. First determine the coordinates of each of the corners of the cube measured from the origin of the XYZ axes. When using a hand calculator a computation table should be set up similar to that shown in Section 18.3 for isometric. When using a digital computer a simple program can be written as shown in Table 18.4.

The resulting perspective drawing is shown in Figure 18.13.

The perspective drawing in Figure 18.13 has a certain amount of distortion that can be reduced by increasing the Station Point distance D. For example, D was increased to 150 by simply replacing one punched card in the program (D = 86.4 changed to D = 150). The resulting perspective is shown in Figure 18.14.

```
        $JOB     WATFIV
 1               H = 58.4
 2               R = 43.2
 3               D = 86.4
 4      5        READ,X,Y,Z
 5               IF(X.EQ.400)STOP
 6               DEN = D + Y
 7               XL = ((Y*(R-X))/DEN) + X
 8               YL = ((Y*(H-Z))/DEN) + Z
 9      6        FORMAT(' ',5X,F7.2,5X,F7.2,5X,F7.2,5X,F7.2,5X,F7.2,5X,F7.2)
10               PRINT 6,X,Y,Z,XL,YL
11               GO TO 5
12               END
        $ENTRY
        18.03    76.45    33.02    29.85    44.96
        47.75    91.19    33.02    45.42    46.08
        62.48    61.21    33.02    54.48    43.56
        32.77    46.74    33.02    36.42    41.94
        18.03    76.45    0.000    29.85    27.43
        47.75    91.19    0.000    45.42    30.00
        62.48    61.21    0.000    54.48    24.23
        32.77    46.74    0.000    36.42    23.06
```

Table 18.4 Computer Program – Perspective Projection

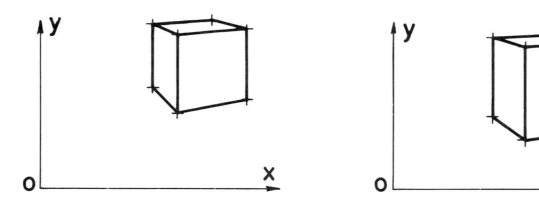

Figure 18.13 Cube Perspective (D = 86.4 mm)

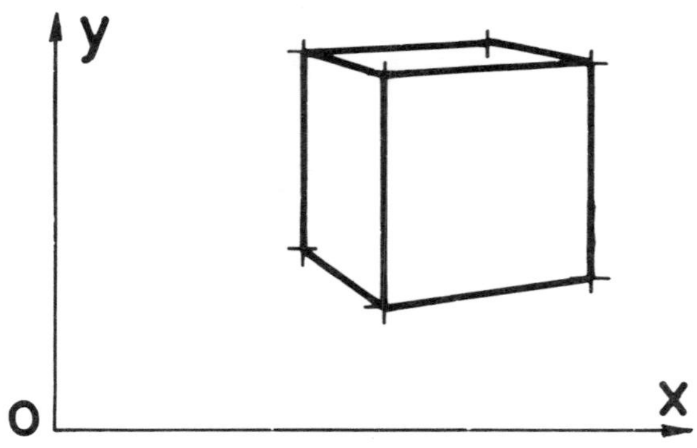

Figure 18.14 Cube Perspective (D = 150 mm)

An example of an object with more detail than the simple cube is shown in Figure 18.15.

A decision has to be made as to the relative location of the house, the picture plane, and the point of sight. In this case the front of the house was placed at 30° to the picture plane. The station point was selected to be at eye level. The only difficulty is keeping track of all the points generated by the computer. A simple numbering of each significant point is useful. The resulting perspective as shown in Figure 18.15 was drawn using 54 separate points.

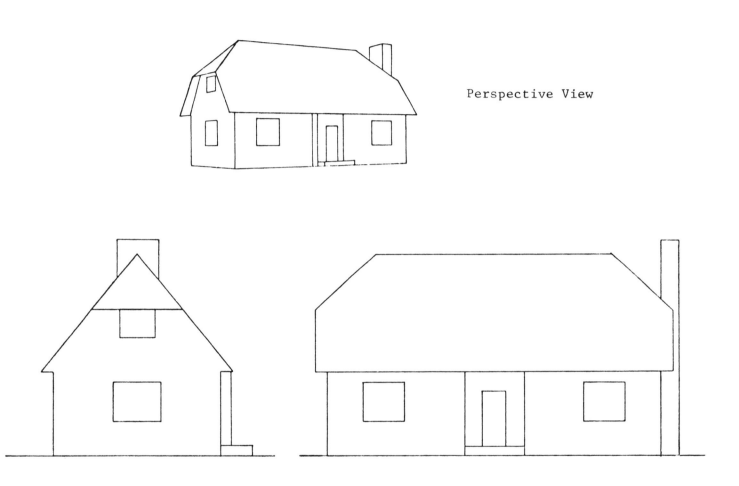

Perspective View

Orthographic Views

Figure 18.15 Perspective Example

PROBLEMS

Dimensions are in millimetres unless otherwise specified.

P 18.1 The top and front views of an object are shown in Figure P 18.1. Obtain the X, Y, and Z coordinates of each corner using the datum axes shown. Produce an isometric view by computing the isometric coordinates either by hand calculator or digital computer. Plot by hand (or by computer plotter) and join the correct coordinate points to produce the isometric drawing.

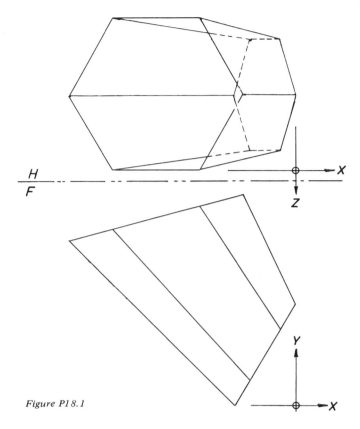

Figure P18.1

P 18.2 The outline of the church shown in Figure P 18.2 can be represented by 30 separate points (and lines) in space. Obtain the coordinates for these points. Choose a station point about 2 m above the ground and 80 m away from the picture plane. Compute the perspective points. Plot and join appropriate points to form the perspective. Now change the station point, for example move back to 120 m and draw another view. Then increase the height to 20 m and obtain another view. Change the position of the picture plane and note the change in the perspective view.

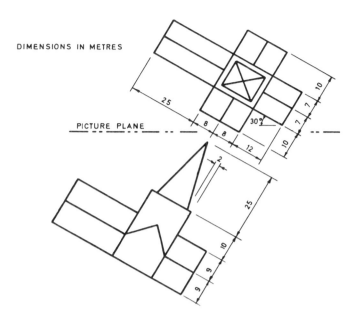

Figure P18.2

P 18.3 Sculptured surfaces, such as the one shown in Figure P 18.3(b) may be produced on a numerically controlled machine tool. The tape instructions for the machine tool may be checked on a numerically controlled drafting machine to produce the three conventional views shown in Figure P 18.3(a) (top, front, and profile). However, these views may not provide a quick visual impression of the surface. An example of the tool path for a surface is shown in the three standard views of Figures P 18.3(a). Produce a perspective view of the surface. It is suggested that five points on each horizontal pass may be used to save time. Compute the coordinates of the perspective view using a station point 200 mm in front, 200 mm above and 100 mm to the left of the left-hand corner. Plot and use a french curve to complete the view as in Fig. P 18.3(b).

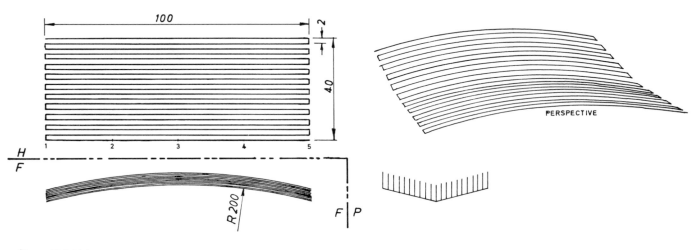

Figure P18.3(a)   *Figure P18.3(b)*

P 18.4 The top and front orthographic views of a simplified space research vehicle are shown in Figure P 18.4(a). It is required to show how the vehicle might appear as it approaches a stationary camera which has its sight point 100 m above the vehicle and is effectively 200 m from the picture plane. Compute the perspective coordinates and draw the perspective view of the vehicle in the four positions shown. Figure P 18.4(b).

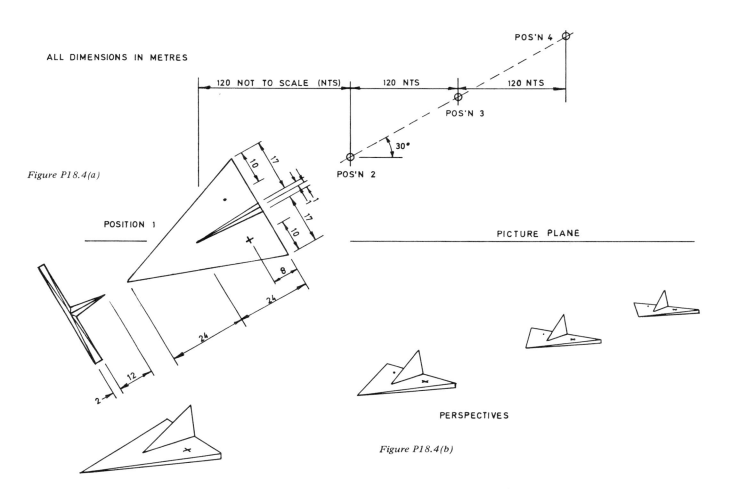

Figure P18.4(a)

Figure P18.4(b)

19. Intersections and Developments

Intersections

Most objects of an engineering nature consist of a collection of regular shapes such as cubes, cones, cylinders, prisms, etc. Two of these shapes will meet in a line of intersection. The location of this line is found by joining a number of separate points. Each point must be carefully located so that it is common to both surfaces. The line of intersection is required to produce a finished drawing and to make a development.

There are two methods in common use to find the line of intersection. Both methods involve the location of the point where a line on one surface pierces the other surface.

19.1 EDGE-VIEW METHOD

This method is useful when one or both of the surfaces is a plane. It involves using the edge view of one of the intersecting surfaces. The method is illustrated in Figure 19.1 in which we require the intersection of the pyramid ABCD and the prism EFGHJKLM. The intersection is seen clearly in Figure 19.1 but our problem is to find the intersection given the horizontal and frontal projections of the solids as shown in Figure 19.2(a).

Procedure

Remember that we wish to find the points where lines on one surface pierce the other surface. Consider the edges AB and AC. The points where these intersect the surface FGKL may be seen in the horizontal view in Figure 19.2(b) since FGKL appears there as an edge view. Project to the frontal view to fix their location, points 1 and 2. In the same manner line AD intersects the surface HGLM at point 3.

We now need to find where the edge GL of the prism intersects the surfaces ACD and ABD of the pyramid. To do this we require an edge view of these surfaces. This may be done by constructing an auxiliary view showing the point view of the common line AD of these two planes. Recall that a point view of a line may be obtained by taking a view in the direction looking along the true length of the line. In this case $a_H d_H$ is a true length; therefore choose an auxiliary plane 1 perpendicular to $a_H d_H$ and complete the auxiliary view, Figure 19.2(b). From the auxiliary view it can be seen that the edge $g_1 l_1$ intersects the surfaces $a_1 c_1 d_1$ and $a_1 b_1 d_1$ at the points 4_1 and 5_1 respectively. These points are transferred back to the horizontal and frontal projections.

The lines of intersection can now be completed by joining the piercing points 1, 2, 3, 4, and 5 as shown in Figure 19.2(b). The visibility of the lines in each view must be determined as usual.

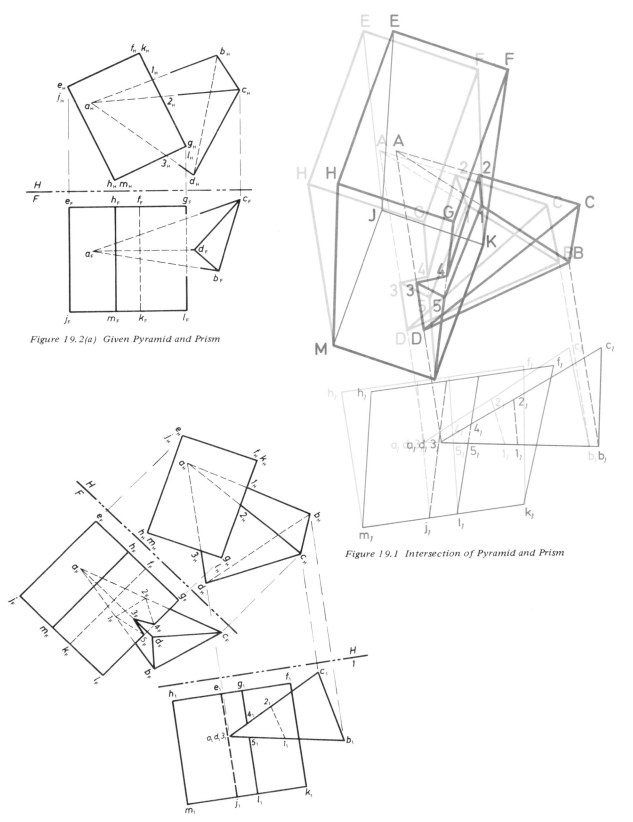

Figure 19.2(a) Given Pyramid and Prism

Figure 19.1 Intersection of Pyramid and Prism

Figure 19.2(b) Construction of Intersection of Pyramid and Prism

19.2 CUTTING-PLANE METHOD

In this method a series of cutting planes are passed through both solids. The planes are chosen so that they will cut known sections such as circles, straight lines, or rectangles, from each of the two solid objects. For example a plane perpendicular to the axis of a cone will cut a circle and a plane parallel to the axis of a cylinder will cut a rectangle. Cutting planes may be horizontal, vertical, or inclined. The choice of which to use depends on the particular problem to be solved. In addition cutting planes should be passed through key points of the intersection. Key points are those points on the line of intersection where the direction of this lines changes. For plane surfaces key points are points that limit the position of the intersecting line. A key point on a curved surface is where the intersecting line is tangent to either one of the two surfaces which intersect.

Horizontal Cutting Planes

As an example we will find the line of intersection of the cone and cylinder given in Figure 19.4. Since the cone is upright and the cylinder is horizontal, it can be seen that a series of horizontal cutting planes will cut circles from the cone and rectangles from the cylinder, Figure 19.3(a).

Procedure

Pass a series of horizontal cutting planes through the cylinder and cone. In order to locate each individual plane draw a profile view of the cylinder as shown in Figure 19.4. Identify the planes and intersection points by a numbering system. It is convenient to start this system on the profile view as indicated by the numbers 1 to 12. Each plane is projected across to the frontal view and transferred up to the right-hand end of the cylinder in the top view. The only key points will be at the top and bottom of the cylinder. These may be found by inspection and are the points 1 and 7.

Consider the plane CC. This cuts the cone to produce a circle of diameter A, while at the same time it cuts the cylinder to produce a rectangle of width B. This circle and rectangle intersect at points X and Y. The detailed procedure follows; refer to Figure 19.4. On the profile view choose a horizontal plane CC and locate points x_P and y_P. This plane determines the width B of the rectangle and the diameter A of the circle as shown in the profile and frontal views respectively. In the horizontal view draw a circle with centre at the apex of the cone and diameter equal to A. Also draw a rectangle of width B, the middle of the rectangle being the top of the cylinder. This circle and this rectangle will intersect at the points x_H and y_H. Project down to locate x_F and y_F. The points X and Y are two points on the required line of intersection.

Repeat the plane, circle, and rectangle procedure outlined above for a series of horizontal cutting planes until enough points have been produced to provide a smooth intersection curve. The complete curve is shown in Figure 19.4 and in Figure 19.3(b).

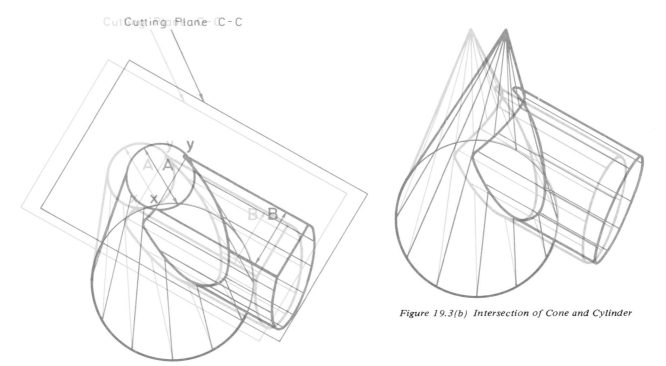

Figure 19.3(b) Intersection of Cone and Cylinder

Figure 19.3(a) Horizontal Cutting Plane

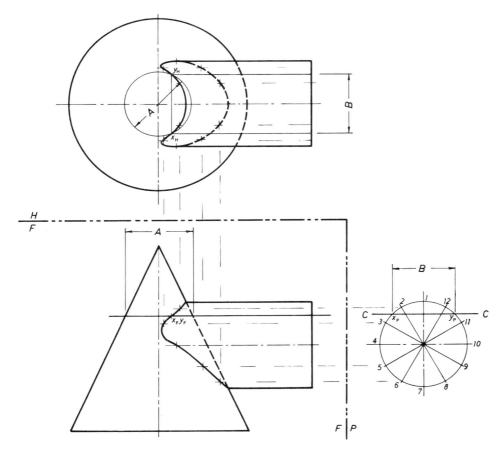

Figure 19.4 Construction of Intersection of Cone and Cylinder

19.3 INCLINED CUTTING PLANES

In general the cutting-plane method makes use of planes selected so as to cut known figures from the intersecting solids. In some cases neither a horizontal nor a vertical cutting plane will do this. An inclined cutting plane may produce a solution.

Consider the intersection of the oblique circular cylinder and the oblique cone shown in Figure 19.5. If a cutting plane is passed through the apex of the cone and parallel to the axis of the cylinder it will cut triangular shapes from the cone and rectangular shapes from the cylinder. The intersecting edges of these shapes will establish points on the intersection curve. A series of similar cutting planes are used to produce enough points so that the curve can be drawn.

Procedure

Draw line AB through the apex of the cone and parallel to the cylinder axis. Let this line terminate at point B where it pierces the horizontal plane containing the base of both cylinder and cone.

Consider one cutting plane as an example. Draw the horizontal intersection of the plane $b_H 1_H 2_H 3_H 4_H$ in Figure 19.6. This plane will cut triangle A12 and rectangle CD34. These will intersect to establish points $e_H f_H g_H h_H$ on the required curve. The frontal projection is found by the normal procedure of projecting down to locate 1_F, 2_F, 3_F, 4_F, c_F, and d_F. Join points as shown in Figure 19.6 to find the projections e_F, f_F, g_F, h_F.

Select another cutting plane and repeat above procedure. When sufficient points have been obtained, join them and take visibility into account. One limiting plane is at the rear tangent to the cylinder and one is at the front tangent to the cone as shown in Figure 19.6 and 19.7.

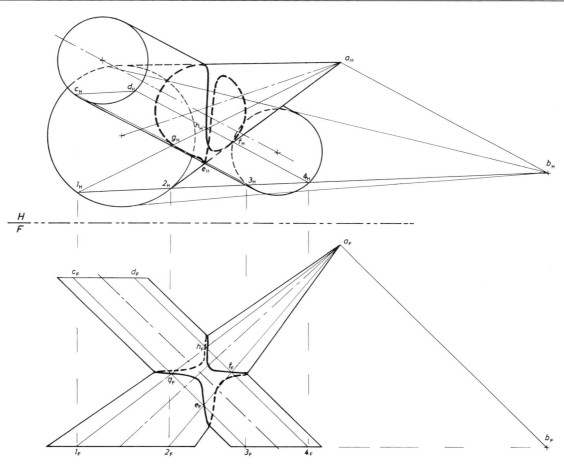

Figure 19.6 Intersection of Cone and Cylinder

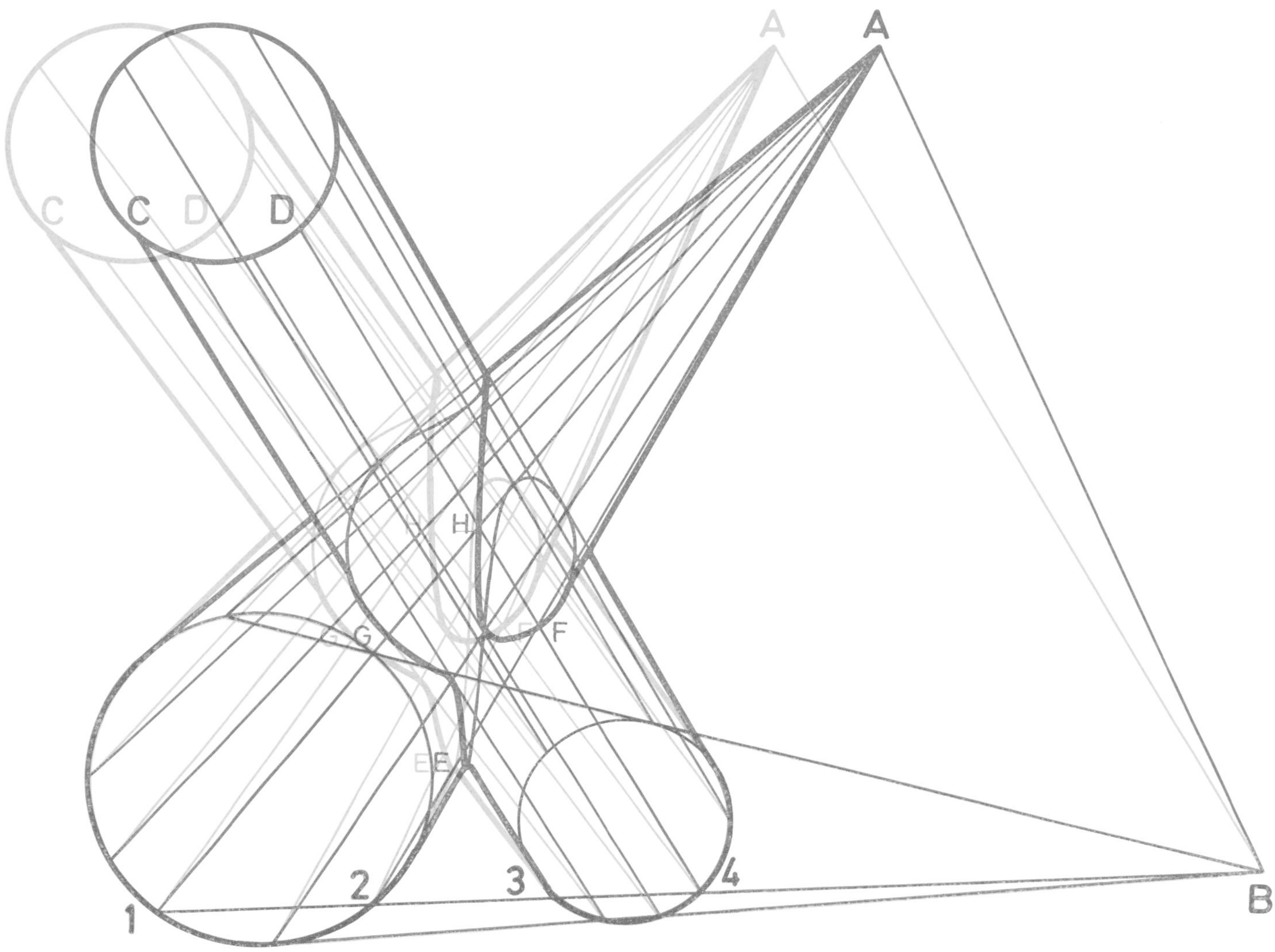

Figure 19.5 Oblique Cone and Oblique Cylinder

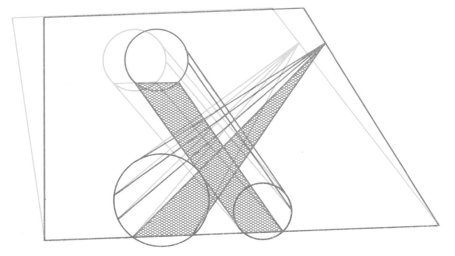

Figure 19.7 Intersection of Oblique Cone and Oblique Cylinder

19.4 DEVELOPMENTS

A great many manufactured parts are made from flat sheets of material which are cut and then formed by bending, folding, cutting, or stretching to the required shape. An article to be made in this manner is first drawn as usual in the required orthographic views. Next a development is made and used as a pattern to which the flat sheet is cut before forming. The flat development may also be called a pattern, stretch-out, or blank. The development of an open rectangular box and a hollow circular cylinder are illustrated pictorally in Figure 19.8.

Developments must be accurately drawn so that the correct shape is obtained after the part has been formed. Allowance must be made for bends. This allowance is given in various handbooks and is beyond the scope of this text. The development drawing must be carefully transferred to the flat plate of material. Photographic methods are sometimes used even for very large developments.

Plane and single curved surfaces may be developed. Cylinders and cones are single curved surfaces. Warped and double curved surfaces cannot be developed exactly. That is to say they cannot be made from flat sheets without an additional stretching operation. An example of a warped surface is a hyperboloid, Figure 11.4. A sphere or globe is an example of a double curved surface. However, both warped and double curved surfaces may be approximated so that a development may be made.

The development of a surface is a pattern such that all lines on it are true lengths. The construction of the development therefore involves the determination of the true lengths of all significant lines on the surfaces.

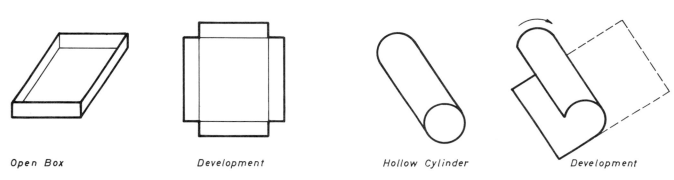

Open Box   Development   Hollow Cylinder   Development

Figure 19.8 Development of a Box and a Cylinder

19.5 DEVELOPMENT OF A CYLINDER

The horizontal and frontal projections of a hollow cylinder are shown in Figure 19.9(a). The development is required.

Procedure

The surface of the cylinder is simply unrolled along a stretchout line. The length of this line is the perimeter of a right section of the cylinder.

As a first step, divide the cylinder into a number of vertical elements. In this example we will choose twelve labelled 1 to 12 as shown. The true length of each element will appear in the frontal view. The true distance d, between each element will appear in the horizontal view. The number of divisions is chosen so that the chordal distance d and the arc distance will be approximately equal.

Construct a stretch-out line of length equal to the circumference of a right section of the cylinder. This can be done by calculation (circumference = π x diameter) and then dividing this line into twelve equal spaces as shown in Figure 19.9(b). Another method of constructing the stretch-out line is by stepping off the distance d with dividers. Project the true lengths from the frontal view to the appropriate line elements on the development. Join these points to produce the complete development.

Note that the development begins and ends on the shortest seam line. This is done so that when the part is rolled to its final shape the joining of the ends at the seam will require the least amount of welding, soldering, or glueing etc. Another important point to note is that the development is laid out with the inside surface uppermost. That is, the pattern must always be folded so that the side nearest the draughtsman is the inner side of the object.

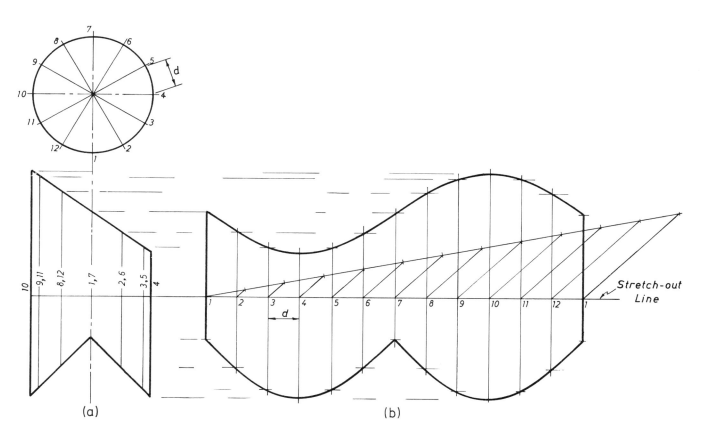

Figure 19.9 Development of a Cylinder

19.6 DEVELOPMENT OF A CONE

The development of a cone is somewhat more involved than that of a cylinder because the true lengths of all the elements are not immediately available. The length of the stretch-out line is the circumference of a right section of the cone. The stretch-out line, however, is an arc of radius equal to the slant height of the cone. The true lengths of all elements are spaced at their true distance apart along the stretch-out line.

Given the horizontal and frontal projections of the truncated cone in Figure 19.10(a), the development will be drawn as an illustrative example.

Procedure

Locate the vertex A of the cone.

Draw a right section through the cone at any convenient position. In this case draw the section through the point B. This section will appear circular in the top view. Divide this circular section up into a series of equally spaced elements, say 12, numbered 1 to 12 in the top view. Project to the frontal view locating the 12 points on the right section through B.

Since the elements on the cone meet at a point, similarly the elements of the development will meet at a point. Therefore choose a suitable point, P, for the vertex of the development and draw the stretch-out line. This will be an arc of a circle of radius equal to length $a_F b_F$. Start with a horizontal element on the development and label this point 1. Step off the distance d with dividers to locate the remaining elements and number them accordingly.

The true length of each element may be obtained by the rotation method. Using this method the true length of an element will appear on the edge $a_F e_F$ by projecting horizontally as shown. For example the true length of both elements 5 and 9 are given by the length $a_F f_F$. Each true length so found is then transferred to the development and marked off on its appropriate element on the pattern. Join the points to produce the complete development as shown in Figure 19.10(b).

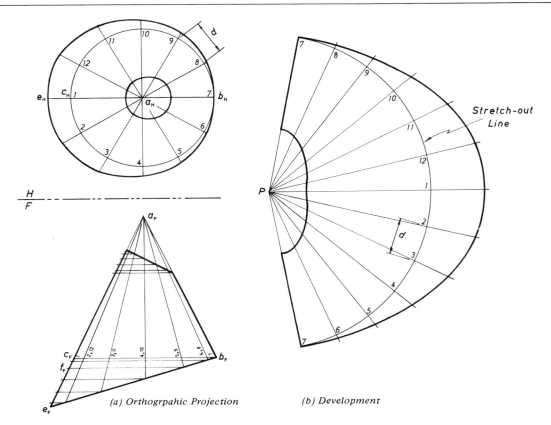

(a) Orthogrpahic Projection   (b) Development

Figure 19.10 Development of a Cone

19.7 DEVELOPMENT OF ELBOW JOINTS

Elbow joints of uniform diameter are designed so that one pattern may be used for all parts of the bend. By this means the waste of material when cutting is reduced to a minimum. The inner radius of the bend is decided upon and then the angle of the bend is divided into an even number of equal angular sections.

As an example consider the 90° four-piece elbow shown in Figure 19.11.

Procedure

Decide on the minimum radius. Knowing the diameter of the straight pipe locate the centre of the bend at P. Draw the inner and the outer radii.

The number of equal angular sections of the bend is equal to two less than twice the number of pieces in the elbow. Using this rule in this case of four pieces, we will divide the 90° bend into 6 equal 15° angles. Each of the interior pieces requires two of these angles and each end piece requires only one. The resulting four parts are identified by the letters ABCD.

Draw an end view (half will do) and divide the circumference into a suitable number of elements, in this case twelve. Project to the view above. Draw a stretch-out line and divide into twelve spaces. Remember that the length of the stretch-out line is the perimeter of a right section of the bend.

The true lengths may be taken directly off the top view and transferred to the development as shown. Note that the developments for Sections B and D have been shifted from left to right so that the patterns nest and can therefore be cut without waste. This means, though, that the seams are alternated in position around the elbow.

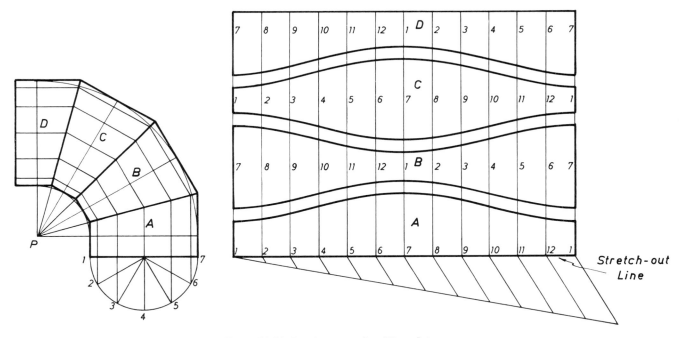

Figure 19.11 Development of an Elbow Joint

19.8 DEVELOPMENT OF A TRANSITION PIECE BY TRIANGULATION

Many parts are combinations of cylinders, cones, prisms, etc. A transition piece is used to connect a section of one shape to that of another shape. As an example consider the transition piece shown in Figure 19.12 which connects a square opening to a circular opening. The surface is a warped surface but it can be developed by dividing it into a series of cones and planes. This results in a series of triangles on the surface. Given the horizontal and frontal views of Figure 19.12, draw the development.

Procedure

Select a seam line P1 as shown. The horizontal view shows the true shape of the openings and the frontal view shows the true length of the seam line.

Divide the circular opening into an equal number of parts, say 12. Join these to the corners, for example, draw 1B, 2B, 3B, 4B, 4C, 5C, etc. This operation produces a series of triangular shapes, J, K, L, and M and four portions of cones, Q, R, S, and T. Each of the four cone parts is divided into four triangles. The sides of these triangles are not shown in true lengths. These sides are the bend lines required to form the part from a flat sheet of material.

To avoid a confusion of lines it is convenient to construct a separate true-length diagram. The true lengths may be found by the rotation method or the short-cut method. Owing to the symmetry of the part the true length diagram is much simplified as shown in Figure 19.12.

To construct the development choose some convenient location for the seam line P1 and with true lengths of 1B and PB as radii, locate point B. Then with true lengths of B2 and chord 12, locate point 2. Continue until the pattern is developed as shown below.

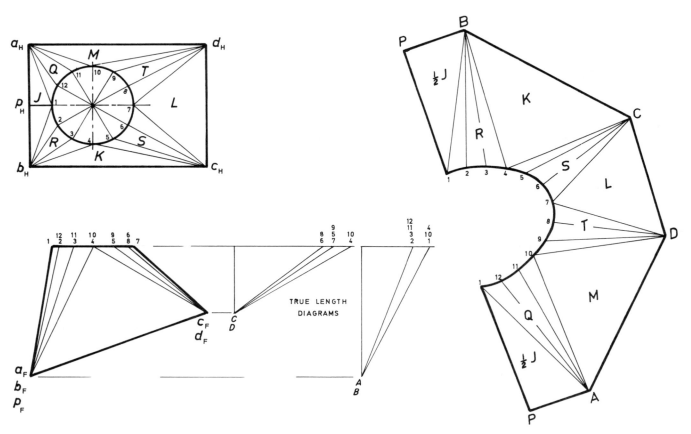

Figure 19.12 Development of a Transition Piece

19.9 DEVELOPMENT OF A WARPED SURFACE BY TRIANGULATION

A warped surface is theoretically non-developable, however any curved surface may be approximated by dividing it into a series of small plane triangles. The smaller the size of the triangles, the closer will the actual surface be approached. Generally, however, surfaces formed in this manner will not be smoothly curved or streamlined, instead they will appear to consist of a series of triangular shapes, which indeed they are. A smoother result may be obtained by developing the surface as a convolute as explained in Section 19.10.

As an example of triangulation consider the development of the transition piece shown in Figure 19.13. This piece connects a circular opening to an elliptical opening. We are given the horizontal and frontal projection and a true-shape view of the circular top.

Procedure

Draw a horizontal plane MM to divide the part symmetrically. Divide the circle into 8 equal parts and the ellipse into 8 equal parts. These are numbered 0 to 8 and lettered A to J respectively as shown. Divide the surface into a number of triangles as shown in the horizontal projection. The development consists of finding the true lengths of the sides of these triangles.

Start at the middle line of the development, 8J, since it is in true length in the top view. Draw 8J at a convenient location as shown. Determine the true lengths of lines 8J and 7J and draw the triangle 8J7 on the development. Draw a similar triangle on the other side of 8J.

Determine the true lengths of the sides of the triangle 7JH and construct this triangle adjacent to triangle 87J. Draw a similar triangle on the other

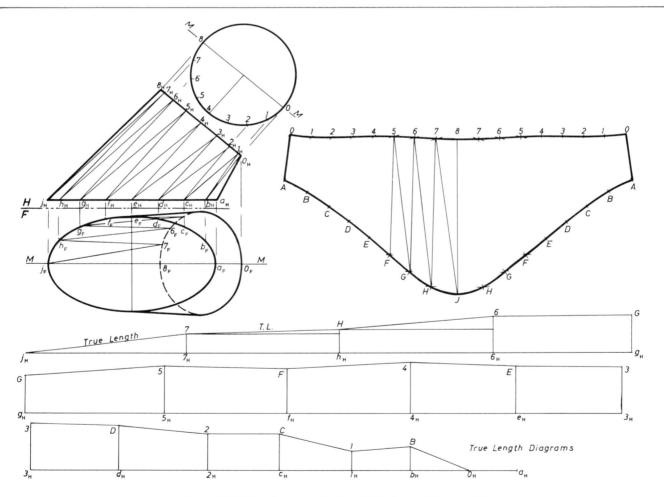

Figure 19.13 Development of a Warped Surface by Triangulation

side of the development. Continue the above procedure and build up the development until complete, as shown.

A convenient method of finding the true lengths of the longer sides of the triangles is shown below the development. Draw a horizontal base line and mark off in succession the horizontal projection lengths $j_H 7_H$, $7_H h_H$, etc. At each point erect a perpendicular of length equal to the distance this point is above the median plane MM. The true lengths are measured at $j_H 7$, 7H, H6, etc., on this construction diagram. The diagram is simply the short cut method of finding true lengths, Section 14.4.

19.10 GEOMETRY OF THE CONVOLUTE
A convolute is a single curved surface like the cylinder and the cone. While the elements of a cylinder are parallel and those of a cone meet at a point, in a convolute two consecutive elements intersect, but no three elements will. The convolute is developable and has some advantages. For example the transition piece of Section 19.9 when made as a convolute will be faired or smoother than if approximated by triangulation. Convolute surfaces are used in aircraft, automobile bodies, ships, transitions, etc.

Geometry
A convolute is formed by elements of a tangent plane. As an example the convolute surface of a certain transition piece will be drawn. The piece in question has end openings of a circle and an ellipse as shown in Figure 19.14.

Procedure—Drawing the Convolute
Locate a number of points on the ellipse, at any convenient spacing, numbered 1 to 12. For each point we require the corresponding point on the circle. Joining each pair of points will produce one of the elements of the convolute. Consider point 9 for example. Draw the tangent P9 to the ellipse. Since the end faces of the transition are parallel, then the tangents to the circle will be parallel to those of the ellipse. The corresponding point 9_c on the circle will be found where line $R9_c$, parallel to P9 is tangent to the circle. Line 99_c is an element of the convolute. Since the radius is perpendicular to the tangent of a circle, it can be noted that the tangents to the circle can be replaced by radii in order to locate points on the circle. Continue the procedure to produce the required number of points to indicate the convolute surface.

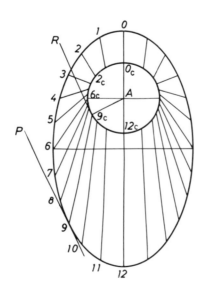
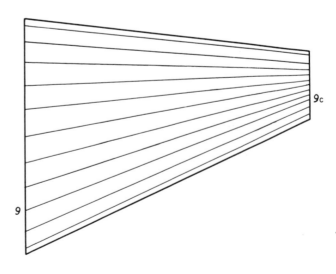

Figure 19.14 Geometry of a Convolute

19.11 DEVELOPMENT OF A CONVOLUTE

The development of a convolute involves finding the true shape of the four-sided area between closely spaced consecutive elements. This can be done by triangulation. Extending two consecutive elements until they meet is useful but these lines meet at such a small angle that their intersection point is difficult to locate accurately. The envelope of the extension of all the elemental lines is a curve (the edge of recession) which may prove useful in detecting errors in construction.

As an example we will consider the design of a convolute to join the circular and elliptical ends of the transition piece shown in Figure 19.15. This piece has the same end shapes as the example of Section 19.9 for purposes of comparison.

Procedure

Given the horizontal and frontal views of the ends of the transition piece, design the convolute joining them and draw its development.

Draw a true-shape view of the circle in an auxiliary view as shown. Determine the intersection of the two planes containing the circle and the ellipse. This is a vertical straight line, the end view being at point X. Draw the frontal and auxiliary views of this line, $x_F y_F$ and $x_1 y_1$.

Divide the ellipse into a number of approximately equal parts, lettered A to J as shown. For each of these points we must find the corresponding point on the circle which will be an element of the convolute. Consider point D for example. Draw a tangent to the ellipse at d_F. This tangent intersects $x_F y_F$ at s_F. Now locate s_1 in plane 1. From s_1 draw a tangent to the circle at 3_1. These two tangents determine a plane tangent to the surface and therefore the line $3_H d_H$ is an element of the required convolute. Continue the above procedure to produce eight elements of the convolute. Note that these are in different locations than the straight lines given by triangulation in Section 19.9.

The next step is to draw the development. In this example we will divide the rectangular strips up into triangles as shown. Proceed to find true lengths and lay out the pattern as detailed in Section 19.9. The result is shown in Figure 19.15.

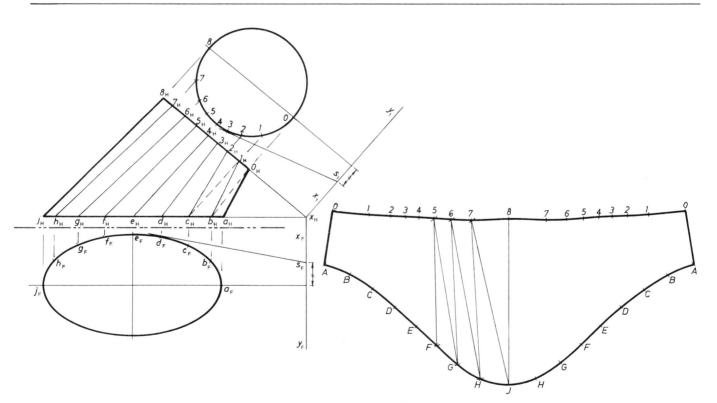

Figure 19.15 Development of a Convolute

19.12 DEVELOPMENT OF A SPHERE

Large spheres are often used as liquid storage tanks. The spherical tank requires less surface material for a given contained volume than any other shape of tank. Since a sphere is a double curved surface the development can only be approximated unless the material is stretched. Two methods of making the development are in common use, the meridian or gore method and the zone method.

Meridian or Gore Development

In this method the sphere is divided into a number of equal gores by meridian planes, Figure 19.16(a). A meridian plane is one that passes through both poles. Each gore is developed by considering it to be part of a cylinder wrapped around the sphere.

Procedure

Given the horizontal and frontal projections of a sphere, divide the top view into an equal number of gores, 16 in this example. In the frontal view divide the circumference into a number of equal lengths, in this case 12. This can be done by drawing a stretch-out line of one gore of length $\pi D/2$, and dividing this into 12 parts numbered as shown. The true lengths of the 12 sections of a gore, ab, cd, etc. are transferred from the frontal view to the development at $a_1 b_1$, $c_1 d_1$, etc. as shown. The remaining 15 gores are laid out identically as indicated in Figure 19.16(b).

This method results in a large number of seams meeting at the poles. Such a situation can be avoided by using a cupped section at the poles, or by using a flat cone as in the zone method.

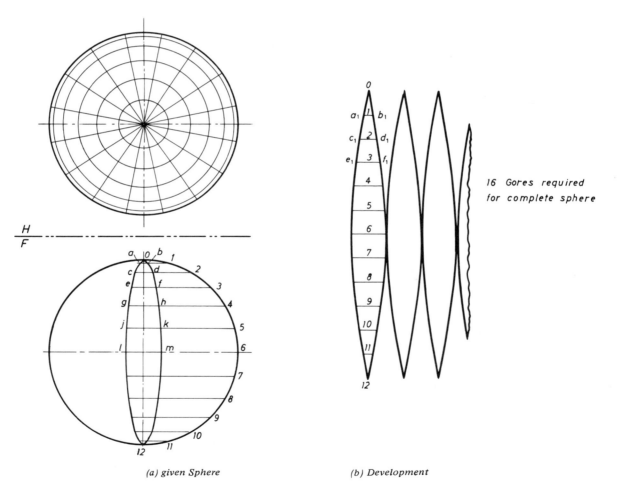

(a) given Sphere   *(b) Development*

Figure 19.16 Development of a Sphere by the Gore Method

Zone Development

In this method the sphere is divided into a number of conical sections. Each conical section is then developed as in Section 19.6. In the example of Figure 19.17(a) the frontal projection is divided into zones by horizontal planes. The polar cap is a right circular cone, the other zones being frustrums of cones.

Procedure

Divide the frontal view into a number of zones, 8 in this case, identified in pairs by the letters A, B, C, D. This is done by the horizontal planes, aa, bb, cc and dd. Zone D is a cone with developed radius of length de. The radii for the development of the other zones are found by drawing tangents to the sphere at b, c, and d and extending them to the vertical centre line. Divide the top view into a number of equal parts, 16 in this example, by radial lines as shown numbered from 0 to 15. Project the bases of the cones up to the top view.

To lay out the development select a point e_1 and draw the circular stretch-out line with radius de. Step off the chords using the chordal lengths from the top view. This produces the development of one polar cap, zone D. To develop zone C use R_2 as radius touching zone D at point d_1. The other radius from the same centre is obtained by adding distance dc to R_2. Step off the distances from the top view to obtain the developed cone base as before. Zones B and A are developed in the same manner as for zone C. Repeat the procedure to obtain the other half of the pattern, Figure 19.17(b).

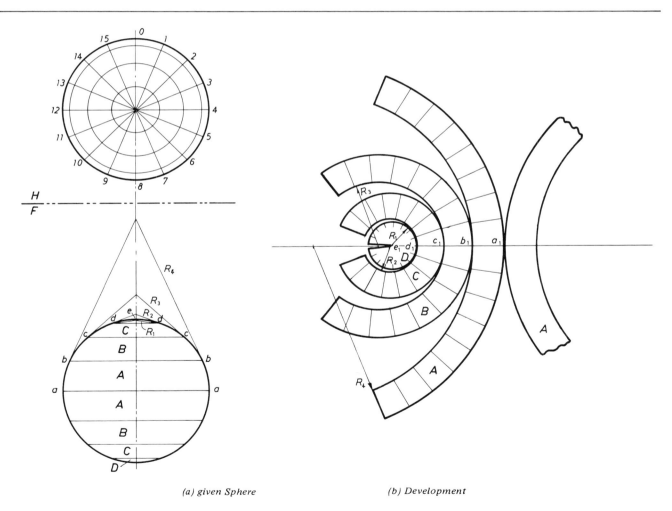

(a) given Sphere   (b) Development

Figure 19.17 Development of a Sphere by the Zone Method

PROBLEMS

P 19.1 Find the intersection of the line AB with the oblique plane CDE. Show correct visibility. Use a vertical cutting plane and then check your results by taking an edge view of the plane. The problem may be drawn from data given in Figure P 19.1(a) or drawn by plotting the points given in Figure P 19.1(b). Locate the H and F views in a suitable position on your drawing paper.

P 19.2 Find the line of intersection of the plane ABC with the prism. Use an edge view of the plane. Lay out from Figure 19.2(a) or by using the coordinates of Figure P 19.2(b).

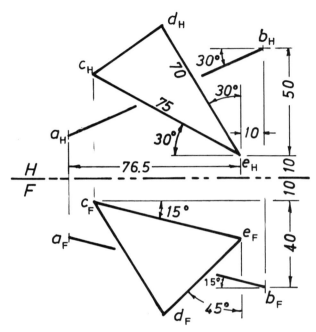

Figure P19.1(a)

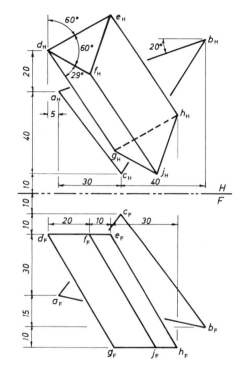

Figure P19.2(a)

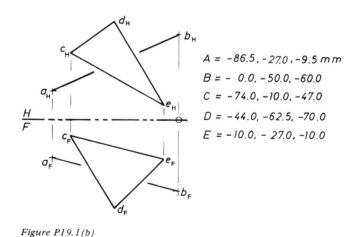

A = −86.5, −27.0, −9.5 mm
B = − 0.0, −50.0, −60.0
C = −74.0, −10.0, −47.0
D = −44.0, −62.5, −70.0
E = −10.0, − 27.0, −10.0

Figure P19.1(b)

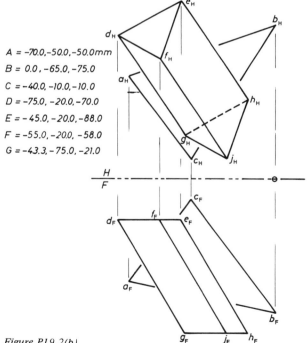

A = −70.0, −50.0, −50.0 mm
B = 0.0, −65.0, −75.0
C = −40.0, −10.0, −10.0
D = −75.0, −20.0, −70.0
E = −45.0, −20.0, −88.0
F = −55.0, −20.0, −58.0
G = −43.3, −75.0, −21.0

Figure P19.2(b)

P 19.3 Find the intersection of the line DE with the right cone. Use an oblique cutting plane and show the correct visibility. Alternate layout method in Figure P 19.3(b).

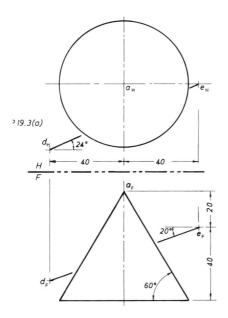

Figure P19.3(a)

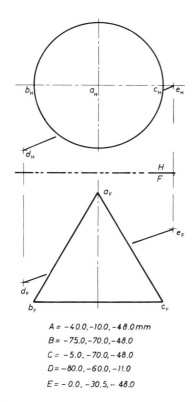

A = -40.0, -10.0, -48.0 mm
B = -75.0, -70.0, -48.0
C = -5.0, -70.0, -48.0
D = -80.0, -60.0, -11.0
E = -0.0, -30.5, -48.0

Figure P19.3(b)

P 19.4 and P 19.5 Find the intersection of the prisms shown. Treat the combination in each problem as one object, i.e., do not show any interior parts.

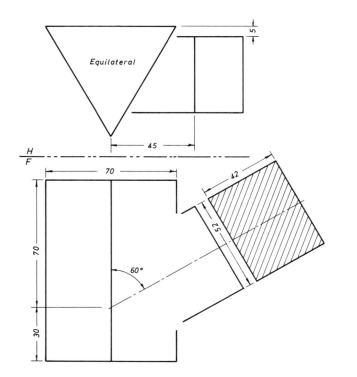

Figure P19.4

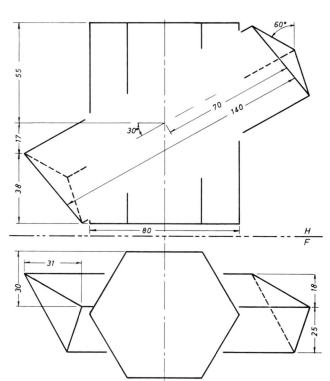

Figure P19.5

P 19.6 and P 19.7 Find the intersection of the cylinders and cones shown. Treat the combination as one solid object, i.e., do not show any interior parts.

P 19.8 A hexagonal bar has been turned, about the offset centre line AA, in a lathe. Find the curve of intersection.

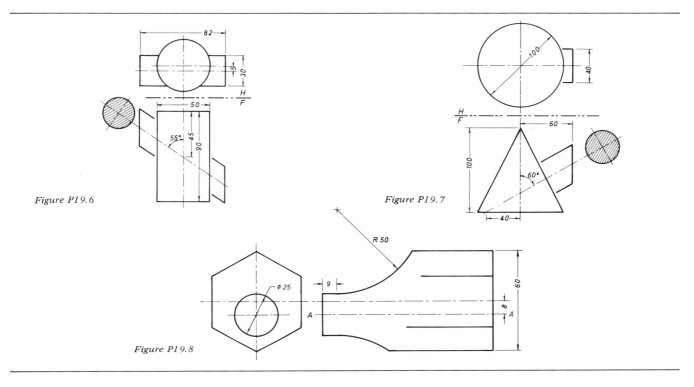

Figure P19.6

Figure P19.7

Figure P19.8

P 19.9 to P 19.11 Draw the developments of the prisms shown. Find the intersections first in Problem P 19.11.

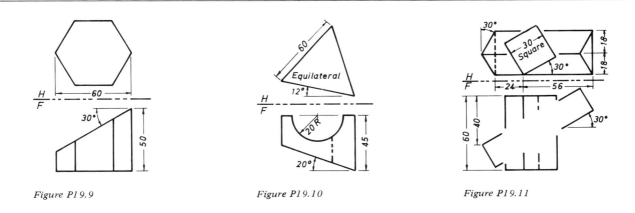

Figure P19.9

Figure P19.10

Figure P19.11

P 19.12 to P 19.14 Draw the developments of the hopper, cylinder, and cone.

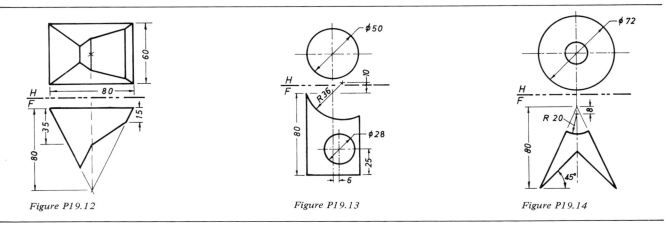

Figure P19.12

Figure P19.13

Figure P19.14

P 19.15 to P 19.17 Draw the developments of the objects shown. Choose a suitable scale for problems P 19.16 and P 19.17.

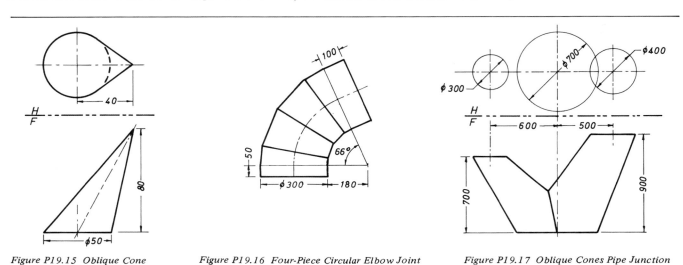

Figure P19.15 Oblique Cone

Figure P19.16 Four-Piece Circular Elbow Joint

Figure P19.17 Oblique Cones Pipe Junction

P 19.18 Draw the development of the cone including the shape of the hole to be cut out for the circular pipe. Find the line of intersection of cone and cylinder.

P 19.19 Draw the development of the fume hood transition piece.

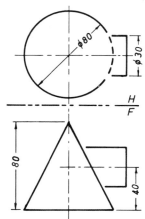

Figure P19.18

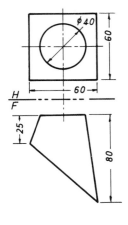

Figure P19.19

P 19.20 Draw the development of the warped surface shown.

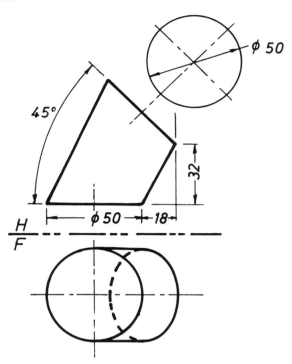

Figure P19.20

P 19.21 Draw a convolute surface joining the two surfaces shown. Lay out the development.

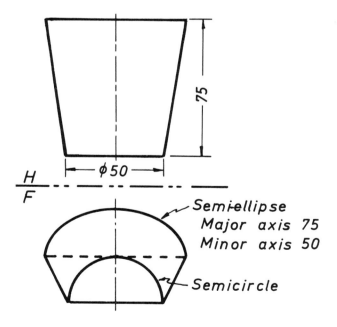

Figure P19.21

P 19.22 Find the intersection of the sphere and the plane ABC. Show the correct visibility. Cutting planes will cut a circle from the sphere and a straight line from the plane.

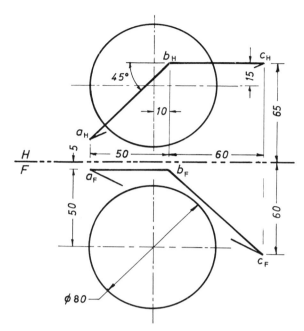

Figure P19.22

P 19.23 Find the line of intersection of the torus and the plane BB. A torus is generated by a circle revolving about an axis lying in the plane of the circle. A torus is shown on page 30.

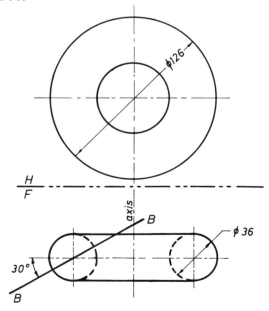

Figure P19.23

20. Interpreting Engineering Drawings

Engineering drawings are one of the most important means of communication between designers and builders of an object or system. Drawings are essential because words alone cannot convey efficiently all the information to be communicated. Engineering drawings are produced using international conventions in such a manner as to avoid misinterpretation as much as possible. These drawings from a universal engineering language made up of lines, symbols, and signs which must be mastered if the information supplied on the drawing is to be understood by those concerned.

This chapter is intended to be an aid to interpreting engineering drawings. It is not designed to instruct you in how to make engineering drawings. It does, however, explain the lines, symbols, and signs so that you can read a technical drawing. First we will consider the "lines" on the drawing, that is the various types of views and sections. Following that we will explain the symbols and signs which form an equally important part of the language of engineering drawing.

20.1 STANDARD VIEWS

The universally accepted standard views on engineering drawings are orthographic projections as explained in Chapters 2, 3, 4, and 5. There are six standard views consisting of: Front, Top, Left Side, Right Side, Bottom, and Rear Views. In addition auxiliary views may be used if necessary to provide further information.

The location of these views on the drawing depends on whether the projection is First-Angle or Third-Angle (see Chapter 3). A symbol should be placed on the drawing to indicate whether First-Angle or Third-Angle projection has been used, Figure 20.1 Third-Angle projection is used in North America, while First-Angle projection is in common use in Europe.

These two projection systems are illustrated in Figure 20.2 which shows a pictorial view of an object and also its First-Angle and Third-Angle projections. It may be noted that the individual views on the two systems are the same. The relative positions of the views, however, are changed as explained in Chapter 3.

20.2 NUMBER OF VIEWS

Those views needed to completely define an object are the only views which should be shown on an engineering drawing. In most cases not all of the six standard views are needed. Three views, top front and side are usually sufficient to show all the detail. In some cases less than three views with suffice.

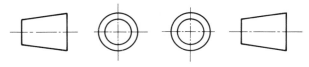

First-Angle Symbol   Third-Angle Symbol

Figure 20.1 Projection Symbols

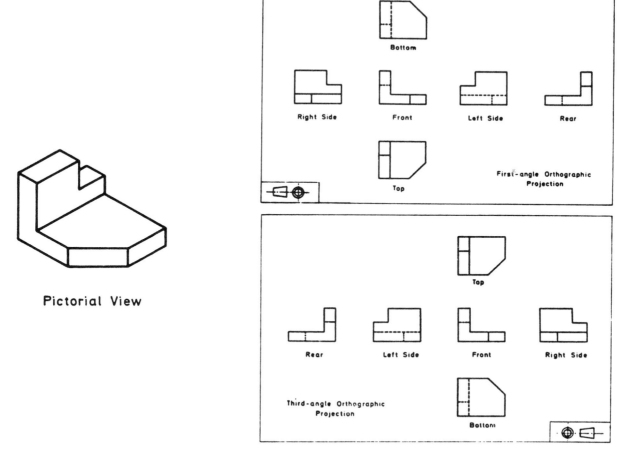

Figure 20.2 Object in First-Angle and Third-Angle Projections

ONE-VIEW DRAWINGS

There are at least two situations in which only one view plus a note is all that is required. One such case involves thin parts where a note is added to define the thickness, Figure 20.3(a). Another situation is the drawing of shafts, bolts, and screws. For these parts notes are used to complete the specification, Figure 20.3(b).

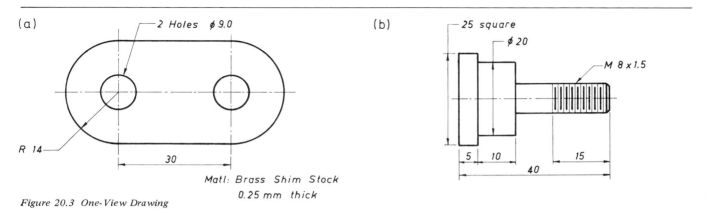

Figure 20.3 One-View Drawing

Two Views

Frequently two views only are required. Two cases are illustrated in the following examples. Figure 20.4(a) shows an object in which the front view and the top view are identical; therefore one of these views should be eliminated. The top view, in this case, should not be presented. In Figure 20.4(b) the side view does not present any more information than that already given in the front and top views. The side view, in this case, should be eliminated as the other two views provide a better representation.

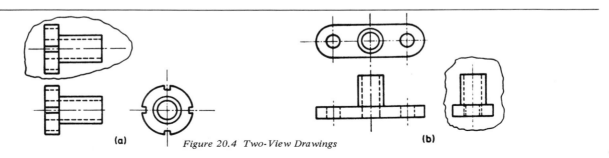

Figure 20.4 Two-View Drawings

Sloped and Curved Surfaces

It is sometimes difficult to recognize sloped or curved surfaces on a drawing as they are not obvious in all views. For example in Figure 20.2, the sloped surface is indicated in the top and bottom views. However, it is not visible as a sloped line in any of the other views. An example of a curved surface is shown in Figure 20.5. Also note that each view provides only two of the three dimensions of height, width, and depth. We need at least two views to show all three dimensions on an orthographic projection.

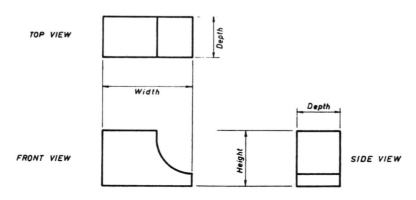

Figure 20.5 Curved Surface

20.3 LINEWORK

Every line on an engineering drawing has a particular purpose. The outline of the object is shown together with the edges only of all significant features of the part. If these lines are visible in any view they are drawn as solid lines. If they are not visible, i.e., hidden, then they are drawn as dashed lines. One drawing convention uses three different thickness of lines: thick, medium, and thin. Another convention uses only two thicknesses of lines. This convention specifies the inked thick line to be approximately 0.7 mm wide and the thin line to be 0.3 mm wide. Various types of lines are shown in Figure 20.6.

LINEWORK

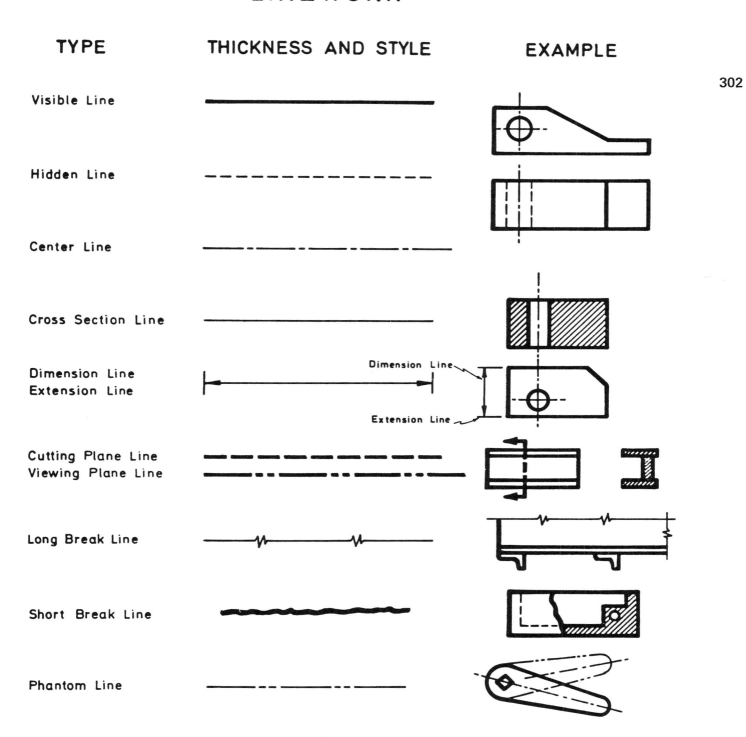

Figure 20.6 Types of Lines

20.4 AUXILIARY VIEWS

In some situations none of the six standard views of Figure 20.2 will show the true geometrical relationship required. Examples are true-shape views and dihedral angles. In this case an auxiliary view is required (Chapter 5).

An example is shown in Figure 20.7. The true shape of the bent portion of the part is shown only on the auxiliary view. The direction in which the view was taken is indicated by the dashed line, letters, and arrows. The view itself may be identified by the words "View A–A". In this example it is a partial view only as the whole part need not be shown in the auxiliary direction.

20.5 SECTIONAL VIEWS

The interior of an object is frequently shown in a sectional view, sometimes called a "section". A section is a view obtained by considering the part to be cut open or sectioned. The location of the cut is shown by a cutting line. As indicated in Figure 20.8, the arrows on the cutting line indicate which part we are looking at after cutting. The face assumed to have been cut is shown by parallel section lines or cross-hatching.

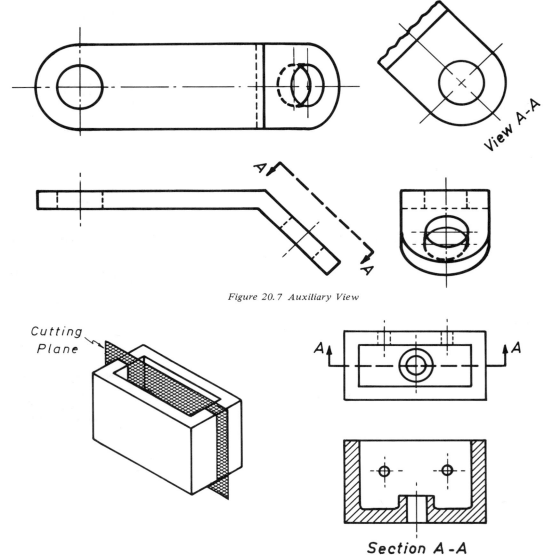

Figure 20.7 Auxiliary View

Figure 20.8 Sectional View

Half-Section

A half-section may be drawn instead of a full section as above. This convention is often used for symmetrical objects. It has the advantage of showing both the exterior and interior of the object in one view, Figure 20.9.

Broken-Out Section

For some objects only a portion of the part may need to be shown in section. In this case a break line is used to bound the broken-out section, Figure 20.10.

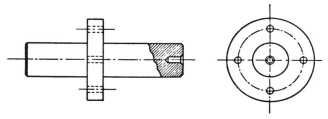

Figure 20.10 Broken-Out Section

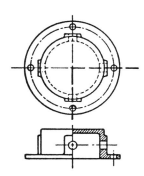

Figure 20.9 Half Section

Revolved Section

These are used to show the sectional shape of an arm, spoke, or similar long object. The section is made by assuming a cutting plane and then revolving the plane through 90°. In Figure 20.11(a) the section has been revolved and in Figure 20.11(b) it has been revolved and broken out from the rest of the view.

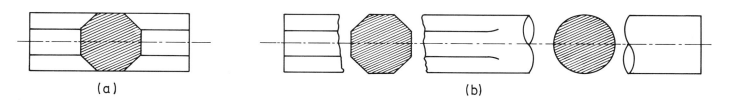

Figure 20.11 Revolved Sections

Offset Section

In order to show the sections of interest of an irregular object, the cutting plane may not be straight but may consist of several parts offset or at an angle to each other. An example is shown in Figure 20.12.

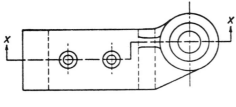

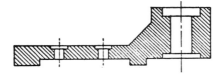

Figure 20.12 Offset Section

Removed Section

A section may be removed from its normal position in the conventional view arrangement. Such a section must be clearly labelled so that its location can be identified without confusion. A number of removed sections are shown in Figure 20.13.

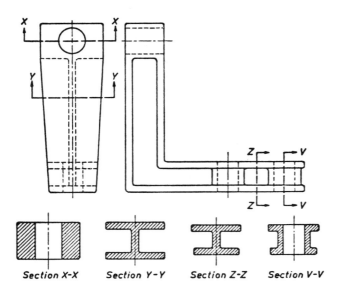

Figure 20.13 Removed Sections

20.6 Dimensioning

Dimensions on a drawing state the actual size of the object; linework indicates the shape. It should not be necessary to use a scale to measure on a drawing or to calculate dimensions by addition or subtraction of other dimensions. There are conventional methods of providing the necessary geometric information and the following information is intended as a guide to interpreting these methods. Additional information is supplied in Section 20.16, Limits and Fits, and Section 20.17, Geometrical Tolerancing.

A simple object is shown dimensioned in Figure 20.14. The part itself is shown with thick lines while the dimension and centre lines are thin. Note the symbol R for radius and ϕ for hole diameter.

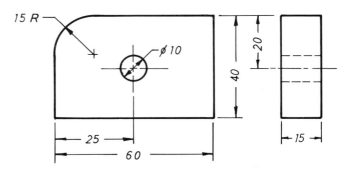

Figure 20.14 Dimensioned Part

It is important that the drawings should be clear and easy to read. For this reason it is preferable not to clutter up the drawing with dimensions. Some simplifications therefore are in common use. For example, if there are a number of holes of the same diameter it is not necessary to dimension completely each hole. Figure 20.15 shows an example where the diameter of all holes is given in a note. Another example is shown in Figure 20.16 for a number of different sized holes.

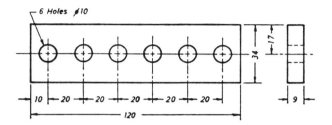

Figure 20.15 Multiple Holes

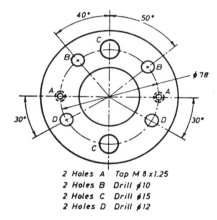

Figure 20.16 Hole Dimensioning

Multiple Parts

A series of parts may have a similar shape but different individual dimensions. Examples are screws, bolts, rivets, etc., one example of which is shown in Figure 20.17. Frequently the dimensions are identified by letters and referred to a table as shown.

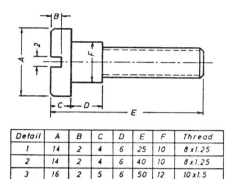

Detail	A	B	C	D	E	F	Thread
1	14	2	4	6	25	10	8 x 1.25
2	14	2	4	6	40	10	8 x 1.25
3	16	2	5	6	50	12	10 x 1.5
4	20	3	6	6	50	12	10 x 1.5

Figure 20.17 Multiple Part Tabular Dimensioning

Datums

A datum is a point, line, or plane on a part from which the position of other features of the object are measured. The datum should be accurately finished and clearly indicated. As an example Figure 20.18 shows a part on which the holes are located from the bottom and side. These are datum surfaces. Note that the holes are not located with respect to each other but are located from the datums. The reasons for the use of datums is because of the end use of the part or because of the method of manufacture.

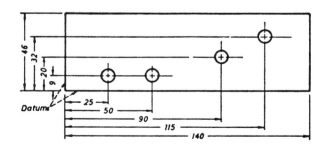

Figure 20.18 Dimensioning from Datums

20.7 Surface Finish Symbols

A part which is made by manufacturing operations on stock material will have a surface finish which depends on the particular operation it has undergone. For example a hack saw cut will leave a rough finish while grinding can produce a smooth finish. It may be necessary to specify on the drawing the quality of the surface finish required. This is done by putting a surface finish symbol adjacent to the particular area concerned.

There are three characteristics of a machined surface and all these may have to be controlled. These are:

1) *roughness* — refers to the peaks and valleys on the surface
2) *waviness* — refers to a longer or larger variation on which the above roughness is superimposed
3) *lay* — refers to the direction of the tool marks.

The standard surface roughness symbol $\sqrt{}$ is applied to a drawing together with adjoining numbers, Figure 20.19. The numbers used in Figure 20.19 are examples only. Different machining processes produce different surface finishes.

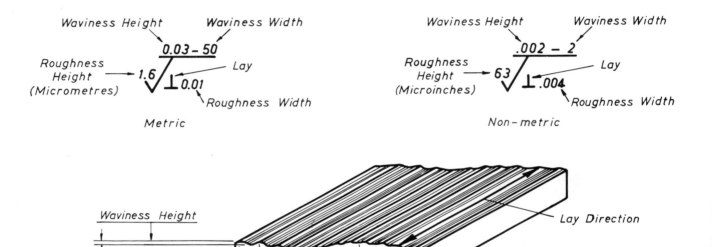

Figure 20.19 Surface Finish Symbol and Explanation

The table below shows approximate values of surface roughness obtainable by various production methods. The ranges shown are approximate because surface finish depends on the skill of the machinist, the condition of the machine and the quality of the cutting tools.

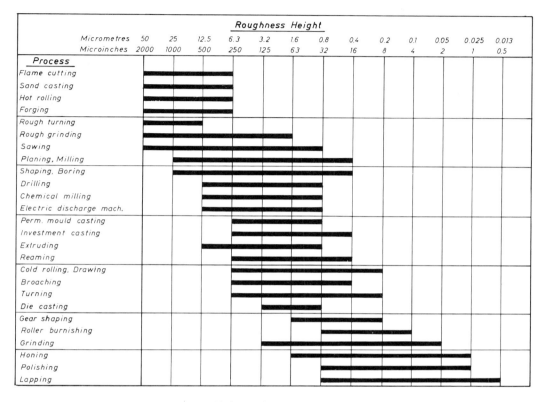

Figure 20.20 Surface Roughness Values

Lay

A cutting tool will leave marks on the surface of the work. These marks will have a certain direction depending on the machining process. In some cases this direction, called lay, is important and may be specified using the symbols shown in Figure 20.21.

LAY SYMBOLS					
Symbol	Explanation		Symbol	Explanation	
∥	Lay parallel to the line representing the surface to which the symbol is applied		X	Lay angular in both directions to line representing the surface to which symbol is applied	
⊥	Lay perpendicular to the line representing the surface to which the symbol is applied		M	Lay multidirectional	
C	Lay approximately circular relative to the center of the surface to which the symbol is applied		R	Lay approximately radial relative to the center of the surface to which the symbol is applied	

Figure 20.21 Symbols for Lay

20.8 Screw Threads

The screw thread is a common type of fastener. It is basically a helical groove around a cylindrical surface. The example shown in Figure 20.22 has a vee shape. Power screws may have a square or trapezoidal thread shape.

The thread is left or right hand depending on the direction of the helix. A right-hand screw will advance into a surface when it is turned clockwise as you face the surface. More than one thread may be cut side by side on the cylinder. These are known as multiple start threads.

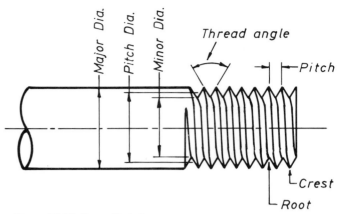

Figure 20.22 Screw Thread

Conventional Representation

Screw fasteners are used so frequently that it would be too time consuming to draw them as they actually appear. Standard simplifications have therefore been developed. There are two such symbols for screw threads, "regular", Figure 20.23(a) and "simplified", Fig. 20.23(b).

Figure 20.23 Screw Thread Representation

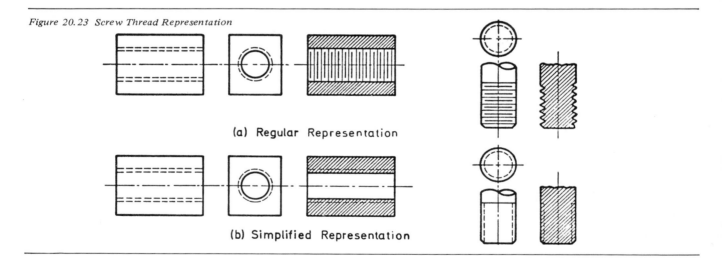

(a) Regular Representation

(b) Simplified Representation

Metric and Unified (Non-Metric) Threads

There are fine and coarse threads in both systems. For example, in the non-metric system there is a 1/4" bolt with 28 threads per inch (fine) or with 20 threads per inch (coarse). This information is not shown on the conventional representation, therefore it must be shown by a note on the drawing, Figure 20.24(a). In the metric system the pitch of the thread is given instead of the number of threads per unit length, Figure 20.24(b).

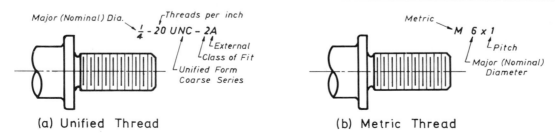

(a) Unified Thread

(b) Metric Thread

Figure 20.24 Screw Thread Notes

Metric and Unified (Non-Metric) Thread Series

The table below shows the metric and unified thread series in common use. The diameters increase as you go down the table. Also the diameter is approximately the same across any horizontal line in the table.

Metric (M) ISO							Unified (UN)			
Basic Major Diameter			Coarse	Fine Thread			Major	Coarse	Fine	Extra F.
Choice 1 mm.	Choice 2 mm.	Choice 3 mm.	Pitch mm.	Pitch mm.			Dia. inches	Threads per inch		
1.6	-	-	0.35	0.2			#0 (.060)	-	80	-
-	1.8	-	0.35	0.2			#1 (.073)	64	72	-
2.0	-	-	0.4	0.25						
-	2.2	-	0.45	0.25			#2 (.086)	56	64	-
2.5	-	-	0.45	0.35			#3 (.099)	48	56	-
3	-	-	0.5	0.35			#4 (.112)	40	48	-
-	3.5	-	0.6	0.35			#5 (.125)	40	44	-
4	-	-	0.7	0.5			#6 (.138)	32	40	-
-	4.5	-	0.75	0.5			#8 (.164)	32	36	-
5	-	-	0.8	0.5			#10 (.190)	24	32	-
-	-	5.5	-	0.5			#12 (.216)	24	28	32
6	-	-	1	0.75			1/4	20	28	32
-	-	7	1	0.75						
8	-	-	1.25	1	0.75		5/16	18	24	32
-	-	9	1.25	1	0.75		3/8	16	24	32
10	-	-	1.5	1.25	1	0.75				
-	-	11	1.5	1	0.75		7/16	14	20	28
12	-	-	1.75	1.5	1.25	1	1/2	13	20	28
-	14	-	2	1.5	1.25	1	9/16	12	18	24
-	-	15	-	-	1.5	1				
16	-	-	2	1.5	1.5	1	5/8	11	18	24
-	-	17	-	-	1.5	1				
-	18	-	2.5	2	1.5	1	3/4	10	16	20
20	-	-	2.5	2	1.5	1				
-	22	-	2.5	2	1.5	1	7/8	9	14	20
24	-	-	3	2	1.5	1	1	8	12	20

Figure 20.25 Thread Series

20.9 Welding

Welding is the process of joining two pieces of metal by raising the temperature of the parts to be united so that they become molten. Usually metal is added in the process. There are many variations in the form of joints to be welded and a particular weld will depend partly on the design of the parts to be joined. Some typical welds are shown in Figure 20.26.

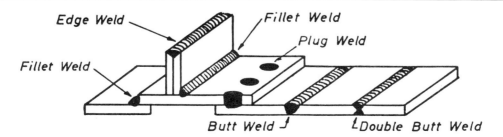

Figure 20.26 Some Types of Weld

On an engineering drawing the location and type of weld are indicated by an arrow symbol as shown in Figure 20.27. The arrow connects the reference line to the side of the joint called the "arrow side". The other or remote side of the joint is called the "other side". Various numbers and symbols are placed around the arrow symbol, some of which are shown in Figure 20.27.

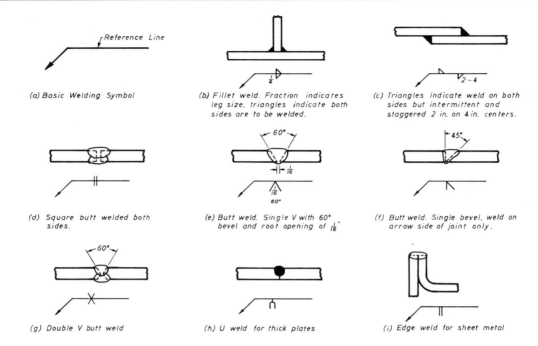

Figure 20.27 Some Welded Joints and Drawing Symbols

U.S. Standard welding symbols are shown in Figure 20.28. The arc and gas welding symbols are indicated in (a) and the electric resistance welding symbols are indicated in (b) of Figure 20.28.

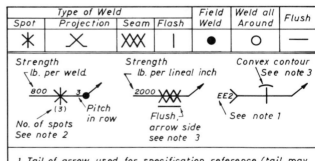

(a) Arc and Gas Welding Symbols

(b) Resistance Welding Symbols

Figure 20.28 US Standard Welding Symbols

20.10 PIPING
Pipes, valves, and fittings are used extensively on many engineering projects. They occur so frequently that they are not usually drawn as projections but are shown as simple linework and symbols. For example the pipework shown in Figure 20.29(a) is usually drawn symbolically as in Figure 20.29(b).

Piping symbols in common use are shown in Figure 20.30.

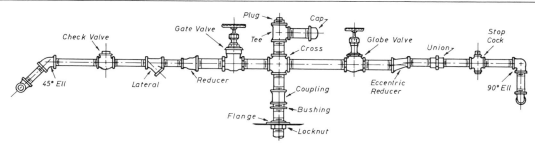

Figure 20.29(a) Pipework—General View

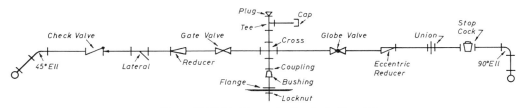

Figure 20.29(b) Pipework—Symbolic Drawing

	Flanged	Screwed	Bell & Spigot	Welded	Soldered
Joint					
Elbow - 90°					
Elbow - 45°					
Elbow - Turned Up					
Elbow - Turned Down					
Elbow - Long Radius					
Reducing Elbow					
Tee					
Tee - Outlet Up					
Tee - Outlet Down					
Side Outlet Tee Outlet Up					
Cross					
Reducer - Concentric					
Reducer - Eccentric					
Lateral					
Gate Valve					
Globe Valve					
Check Valve					
Stop Cock					
Safety Valve					
Expansion Joint					
Union					
Sleeve					
Bushing					

Figure 20.30 US Standard Piping Symbols

20.11 ELECTRICAL AND ELECTRONIC DRAFTING

A large number of symbols are used to represent the variety of electrical and electronic components and sub-assemblies. Further simplification is accomplished by using rectangular or square blocks to represent complete circuits or separate pieces of equipment, Figure 20.31. Single line drawings are used to show the path of the electrical circuit and the individual components, Figure 20.32.

A few of the common symbols are shown in Figure 20.33.

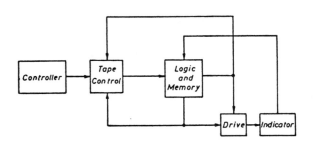

Figure 20.31 Block Diagram

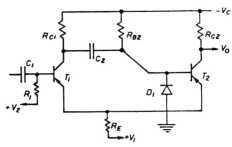

Figure 20.32 Circuit Diagram

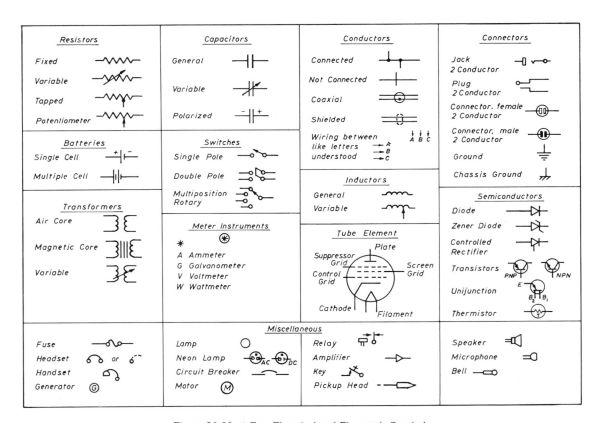

Figure 20.33 A Few Electrical and Electronic Symbols

20.12 STRUCTURAL DRAWINGS

Various symbols and terminology are in standard use on structural drawings. Some of the terms are illustrated in Figure 20.34. An enlarged view of the corner of a truss is shown in Figure 20.35.

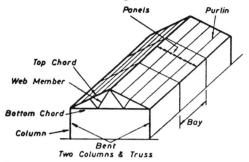

Figure 20.34 Steel Framework

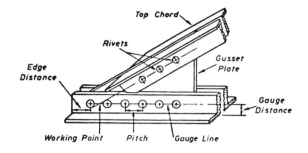

Figure 20.35 Detail of Corner of Truss

The members of a steel structure are usually composed of standard rolled shapes as described below, Figure 20.36.

Item		Order of Data	Abbreviation	Item		Order of Data	Abbreviation
I Beam	I	Depth, symbol, wt./ft., length	12 I 65.0 x 50'-0	Tee	T	Symbol, flange, stem, wt./ft., length	T 3 x 4 x 7 x 19'-5
Equal Angle	L	Symbol, leg, leg, thickness, length	L 6 x 6 x ½ x 12'-9	Structural Tee	⊥	Symbol, depth, cut from, wt./ft., length	ST 5 I x 15 x 12'-3
Unequal Angle	L	Symbol, long leg, short leg, thickness, length	L 3 x 2 x ¾ x 10'-0	Zee	ʃ	Symbol, depth, flange width, wt./ft., length	Z 6 x 3 x 25 x 6'-0
Bulb Angle	⌐	Symbol, web, flange, wt./ft., length	Bulb L 6 x 3 x 16 x 7'-0	Plate	—	Symbol, width in inches, thickness, length in feet	℞ 9 x ⅝ x 2'-10
Wide Flange	I	Depth, symbol, wt./ft., length	12 W 50 x 25'-6	Crane Rail	I	Wt./yd., name, length of run in feet	90 lb A.S.C.E. rail x 50'-0
Channel	[	Depth, symbol, wt./ft., length	10 ⌶ 25 x 20'-0	Bar Stock	■ φ ● φ	Size, symbol, length	1½ φ 5'-10

Figure 20.36 Structural Steel Shapes

The steel members may be assembled using welds, bolts, or rivets. Various types of rivets and bolts are shown in Figure 20.37. Shop rivets and bolts are those driven in the fabricating shop which is usually some distance from the site of the finished structure. Field rivets are those driven on the site at final erection.

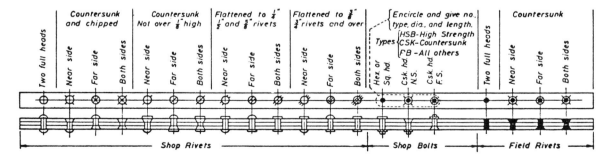

Figure 20.37 Structural Bolts and Rivets

Steel Truss

The drawing used in the design of a truss will show the members as centre lines only as indicated in Figure 20.38. The detailed drawings required for fabrication and erection would be as shown in Figure 20.39. In the case of these detailed drawings two scales are sometimes used. A larger scale is used to show the detail around joints. The other scale is about one-half the detail scale.

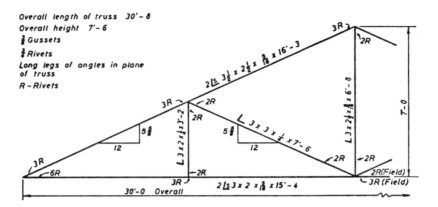

Figure 20.38 Design Drawing

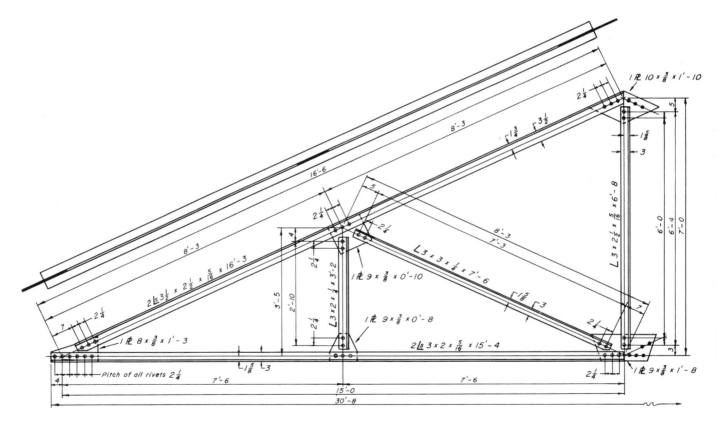

Figure 20.39 Detail Drawing

20.13 ARCHITECTURAL DRAWINGS

There are a large number of symbols used on architectural drawings. There are many symbols for plumbing and electrical fixtures because of the variety of these items. Symbols for wall openings are shown in Figure 20.40. Symbols for commonly used materials are indicated in Figure 20.41.

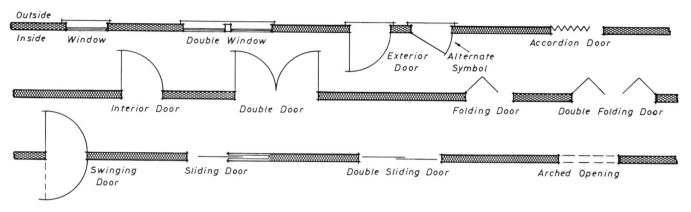

Figure 20.40 Symbols for Wall Openings

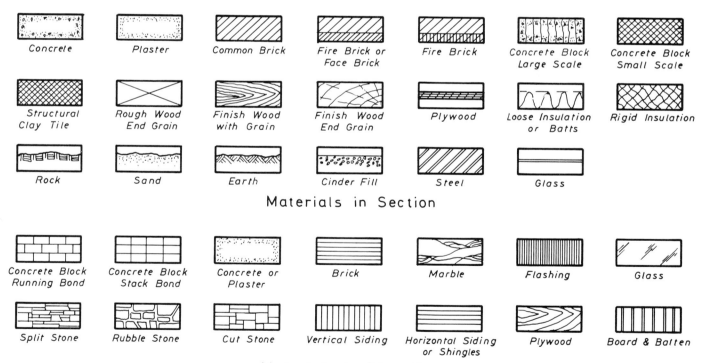

Figure 20.41 Conventions for Materials

20.14 SIMPLIFIED DRAWINGS

In order to reduce the time spent on drawing symmetrical parts and repeated items, simplified drawing conventions are frequently used.

For parts which are symmetrical only one-half or one-quarter of the part may be shown. This is illustrated in Figure 20.42 for a thin gasket symmetrical about the horizontal centre line. In the case where both halves are symmetrical except for a small portion, then the half which contains the deviation is shown together with a note. Figure 20.43. A part having two axes of symmetry may be shown as indicated in Figure 20.44.

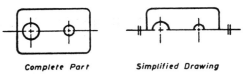

Figure 20.42 Symmetrical Item

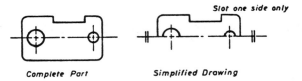

Figure 20.43 Partly Symmetrical Item

Details that are uniformly repeated may be drawn as shown in Figure 20.45. The plate with many holes may show only one hole and the centres of all others.

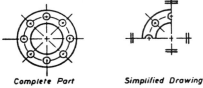

Figure 20.44 Two Axes of Symmetry

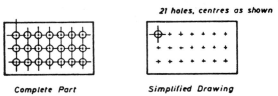

Figure 20.45 Repeated Details

20.15 ASSEMBLY AND SUB-ASSEMBLY DRAWINGS

Many engineering projects consist of a major complex assembly made up of sub-assemblies, each of which in turn consist of a number of separate parts. The individual part may be shown on a detail drawing. A drawing which shows the parts in their correct relation to each other is an assembly or sub-assembly drawing. An assembly drawing should show the relationship of the part clearly and should not be cluttered up with unnecessary detail, particularly hidden portions. Further information is often given in a parts list which is numbered and identified by balloons as shown in Figure 20.46.

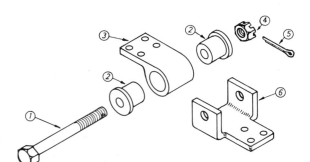

Item	Description	Quantity
①	Bolt $\frac{3}{4}$ Dia. 4" long 1020 Steel	1
②	Bushing, Rubber	2
③	Body, Cast iron	1
④	Nut $\frac{3}{4}$-10 UNC Slotted	1
⑤	Cotter pin $\frac{1}{8} \times 1\frac{1}{2}$	1
⑥	Yoke, 1020 Steel stamping	1

VIBRATION MOUNT

Figure 20.46 Exploded Assembly Drawing

20.16 LIMITS AND FITS OF MECHANICAL PARTS

Limits

In Section 20.6, Dimensioning, the length of a part was indicated by a single value, for example 60 mm. In practice no part can be made to an absolutely exact size. Therefore, limits must be put on the basic dimension to guide the manufacturer as to the precision required. For example, a dimension of 60 ± 0.2 means that the part will be accepted if it measures anywhere between the limits of 59.8 and 60.2 mm.

Fits

When two parts are to be fitted together the drawing must indicate the tightness of the fit required between them. Fits range all the way from parts that have plenty of clearance to those which must be forced together or shrunk on.

The following terms are used to describe limits and fits.

Nominal size is the general size referred to as a matter of convenience. In Figure 20.47 the nominal size is 30 mm.

Actual size is the size obtained by actually measuring a particular part.

Basic size is the size to which the limits refer. See the definitions for Basic Hole System and Basic Shaft System (on the following page).

Limits of size are the minimum and maximum permissible sizes. In Figure 20.47 the limits are 30.05 and 29.98 mm.

Tolerance is the total allowable variation in the size of a part. It is the difference between the upper and lower limit. In the above example the tolerance is 30.05 − 29.98 = 0.07 mm.

Allowance is used in connection with parts which mate together. It is the difference between the upper limit of the shaft and the lower limit of size of the mating hole. An allowance may be positive or negative. For example in Figure 20.48 two mating parts have an allowance of 25.05 − 25.00 = 0.05 mm. There are three types of fit depending on the amount of the allowance.

Clearance fit is one in which the shaft is always smaller than the hole. In Figure 20.48 the allowance will always be positive so that clearance will always exist if the parts are made to the limits shown.

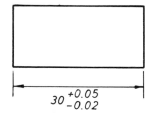

Figure 20.47 Limits on Dimensions

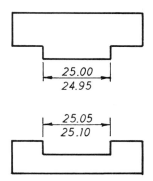

Figure 20.48 Clearance Fit—Mating Parts

Interference fit is one in which the shaft is always larger than the hole. An example is shown in Figure 20.49. The allowance is always negative in an interference fit.

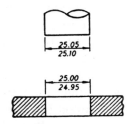

Figure 20.49 Interference Fit

Transition fit. In this case the shaft may be either larger or smaller than the hole. Therefore either a clearance fit or an interference fit may be obtained between the two parts, Figure 20.50.

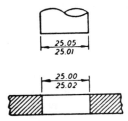

Figure 20.50 Transition Fit

Unilateral and Bilateral Limits

Unilateral limits are designated as those where the minimum and maximum limits of size are placed on the same side of the basic size, Figure 20.51.

Figure 20.51 Unilateral Limits

Bilateral limits are those in which the dimensional limits of size are placed above and below the basic size, Figure 20.52.

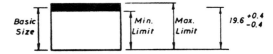

Figure 20.52 Bilateral Limits

Basic Hole and Basic Shaft Systems

There are two bases for a system of limits: the basic hole system and the basic shaft system. The basic hole system is to be preferred as it is more economical than the basic shaft system.

Basic Hole System

This system is used when the shaft can be readily machined to any convenient size and when the hole cannot as it is made by a standard tool such as a drill, reamer, or broach. In this system the minimum hole is considered to be the basic size. An allowance is selected and then tolerances are applied on both sides of this allowance.

The minimum size of the hole in Figure 20.53 is 10 mm and this is the basic size. An allowance of 0.05 is selected and subtracted to determine the maximum shaft size of 9.95 mm. Next the manufacturing tolerance is considered to be 0.05 for the hole and 0.06 for the shaft. The maximum hole size is thus 10 + 0.05 = 10.05 mm. The minimum shaft size is 9.95 − 0.06 = 9.89 mm.

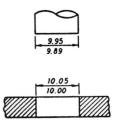

Figure 20.53 Basic Hole System

In the previous example the minimum clearance between the two parts is 10.00 − 9.95 = 0.05 (smallest hole minus largest shaft). The maximum clearance is 10.05 − 9.89 = 0.16 (largest hole minus smallest shaft).

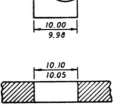

Figure 20.54 Basic Shaft System

Basic Shaft System

The basic shaft system is used when the hole size is to be varied to produce different types of fit around a basic shaft size. A series of drills and reamers is required so that it is more costly than the basic hole system. It may be used where different fits are required at specific places along a long shaft.

An example is shown in Figure 20.54 where the maximum shaft size of 10 mm is the basic size. If the same allowance of 0.05 mm is selected and added to the basic shaft diameter, then the minimum hole is 10.05 mm. Now consider a manufacturing tolerance of 0.05 for the hole and 0.02 for the shaft. The maximum hole becomes 10.10 and the minimum shaft is 9.98 mm.

In this example the minimum clearance between the parts is 10.05 − 10.00 = 0.05 (smallest hole minus largest shaft). The maximum clearance is 10.10 − 9.98 = 0.12 (largest hole minus smallest shaft).

20.17 GEOMETRICAL TOLERANCES

Geometrical tolerances are applied to those situations where it is necessary to control the form or shape of a part more precisely than usual. They are generally used in addition to normal tolerances. In the past these desired relationships were specified by notes added to the drawing such as "surfaces to be square to each other", "surfaces to be parallel to each other", etc.

Geometrical tolerancing should be used for surfaces which come into contact with other parts and when close tolerances are used. However, they should be called for only when necessary to ensure that the design function will be satisfied. This is because their use could increase the cost of manufacture. The symbols refer to the characteristics of attitude, position, orientation, size, and form. Figure 20.55.

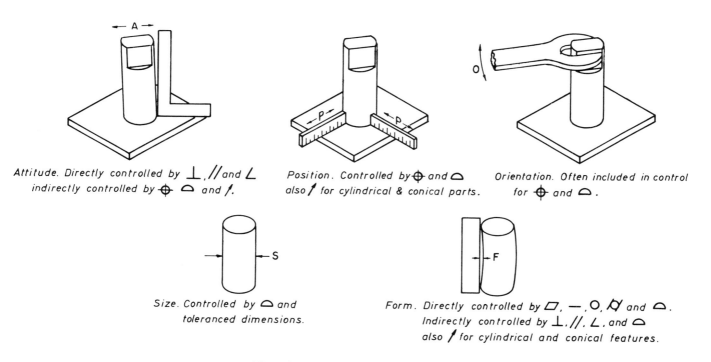

Figure 20.55 Geometrical Characteristics

The symbols indicating geometrical tolerancing agreed upon by the International Standards Organization (ISO) are shown in Figure 20.56.

	Tolerance Characteristics		Symbol	Explanation
Single features	Form tolerances:	Straightness	—	A straight line. The edge or axis of a feature.
		Flatness	▱	A plane surface.
		Roundness	○	The periphery of a circle. The cross-section of a cylinder, sphere, cone or bore.
		Cylindricity	⌭	The combination of straightness, parallelism and roundness of cylindrical surfaces. Mating bores and pistons.
		Profile of a line	⌒	The theoretical or perfect form of a profile defined by true boxed dimensions.
		Profile of a surface	⌓	The theoretical or perfect form of a surface defined by true boxed dimensions.
Related features	Attitude tolerances:	Parallelism	//	Parallelism of a feature related to a datum. It can control flatness when related to a datum.
		Squareness	⊥	Surfaces, lines or axes positioned at right angles to each other.
		Angularity	∠	The angular displacement of surfaces, lines or axes from a datum.
	Composite tolerances	Runout	↗	The position of a fixed point on a surface of a part as it is rotated 360° about its axis.
	Location tolerances	Position	⊕	The deviation of a feature from a true position.
		Concentricity	◎	The relationship between two cylinders or circles having a common axis.
		Symmetry	⌯	The symmetrical position of a feature relative to a datum.

Figure 20.56 Geometrical Tolerance Symbols

Indicating Geometrical Tolerances on Drawings
The geometrical tolerance is shown on a drawing in a small rectangular frame as shown in Figure 20.57.

Note that there is a different order of symbols between the North American and the International conventions at the present time.

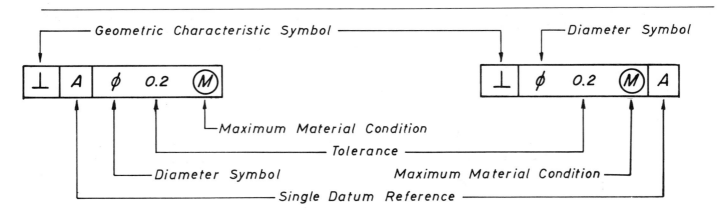

North American Method

International Method

Figure 20.57 Geometrical Tolerance Frame

Maximum Material Condition

Maximum Material Condition (MMC) is indicated by the symbol Ⓜ on a drawing. It means that a feature of a part contains the maximum amount of material permitted by the tolerance. For example an external feature such as a shaft, made to its upper limit would then be at the MMC. For an internal feature, such as a hole, the MMC would occur at the minimum size limit. That is, the hole would have had the minimum amount of material removed during manufacture in order to remain at the MMC. The use of MMC is explained later in Section 20.18.

Datums

A datum is a plane or axis used for manufacture or inspection of a part. It should be clearly identified and accurately finished as measurements are made from it. In some cases three mutually perpendicular planes are used as datums. An example of a datum surface identified by the letter A is shown in Figure 20.58. In this example the datum is the right-hand edge and is subject to a straightness tolerance of 0.04 mm. The leader from the boxed A is connected to the frame by a solid triangle.

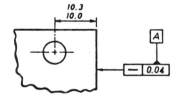

Figure 20.58 Datum Face

When the datum is an axis or a median line of a part then the identifying datum box may be positioned as shown in Figure 20.59.

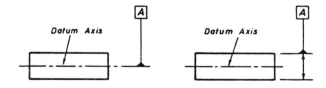

Figure 20.59 Datum Axis

If the datum axis or plane refers to the total length of the part it is shown as indicated in Figure 20.60. When the datum refers to a portion only of the axis it is shown as in Figure 20.61 indicating that only the feature dimensioned is to be the datum.

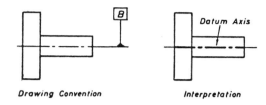

Figure 20.60 Datum is Total Length

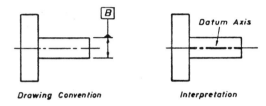

Figure 20.61 Datum is Partial Length

In the situation where either one or the other of two features may be chosen as the datum, the drawing indication is shown in Figure 20.62. This situation may be extended to the case where either of two associated features may be chosen as a datum, B, and in addition another feature, the slot on top, is toleranced from B. Figure 20.63.

A tolerance which is referred to two separate datum surfaces, A and B, is shown by example in Figure 20.64.

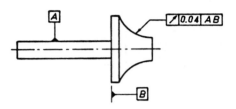

Figure 20.64 Two Datums Specified

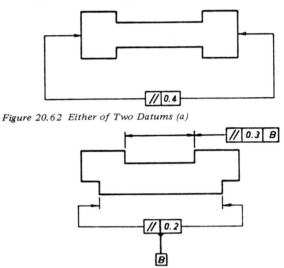

Figure 20.62 Either of Two Datums (a)

Figure 20.63 Either of Two Datums (b)

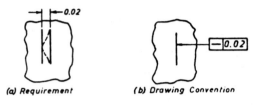

Figure 20.65 Straightness of a Line

20.18 APPLICATIONS OF GEOMETRICAL TOLERANCES BY EXAMPLES

The following examples show how the concept of geometrical tolerances is applied to engineering drawings. The tolerance values given are typical values only; actual values would be specified by the designer.

Straightness

This tolerance controls the straightness of a line on a surface, the straightness of a line in a plane, or the straightness of an axis.

Example 1. A certain line (such as a graduation line on an engraved scale) must lie between two parallel straight lines 0.02 mm apart. The requirement is illustrated in Figure 20.65(a) and the drawing symbol in Figure 20.65(b);

Example 2. The axis of a part must lie in a zone enclosed in a box 0.2 x 0.3 over its entire length. Figure 20.66(a) shows the requirement and Figure 20.66(b) shows the drawing notation.

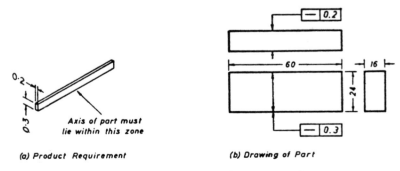

(a) Product Requirement (b) Drawing of Part

Figure 20.66 Axis of Whole Part Toleranced

Example 3. The axis of the part must lie within a cylindrical tolerance zone of 0.04 mm. Figure 20.67.

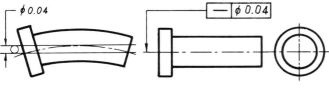

Figure 20.67 *Straightness*

Flatness

This tolerance controls the departure of the surface from a true plane. It specifies the zone between parallel planes. Note that it does not control the squareness or parallelism of the surface to other features. The example in Figure 20.68 indicates that the surface must lie between two parallel planes 0.06 mm apart.

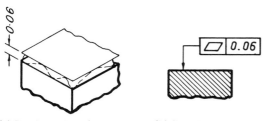

Figure 20.68 *Flatness*

Roundness

This specification is to ensure that any point of a continuous curved surface is equidistant from its centre. The tolerance of roundness requires that the surface lies within a controlled annular zone. It can apply to cylinders, cones, spheres, and any curved surface.

Example 1. The circumference of the round bar must lie between two coplanar concentric circles which are 0.4 mm apart. Figure 20.69.

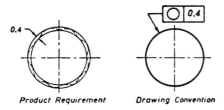

Figure 20.69 *Roundness (a)*

Example 2. This specification by itself can result in a situation where a circular section may not be concentric with the axis of the part and yet the part lies within this tolerance specification. Figure 20.70.

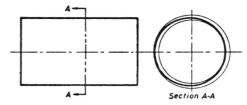

Figure 20.70 *Roundness (b)*

Cylindricity

The concept of cylindricity includes the combination of roundness, straightness, and parallelism applied to a cylinder. The tolerance zone is an annular space between two coaxial cylinders. It is difficult to inspect the combined effects mentioned above, so that each of the conditions of roundness, straightness, and parallelism are often toleranced and checked separately. In the example shown in Figure 20.71, the curved surface of the cylinder must lie within an annular tolerance zone 0.05 mm wide formed by two coaxial cylindrical surfaces.

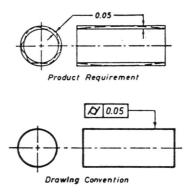

Figure 20.71 *Cylindricity*

Profile of a Line

This tolerance controls the contour. The contour is defined by true position dimensions which are boxed, together with a tolerance zone. Unless otherwise stated, the tolerance zone is considered to be equally spaced about the true form. In Figure 20.72 the profile is required to lie within the bilateral tolerance of 0.4 mm. Boxed true position dimensions are explained later in this chapter.

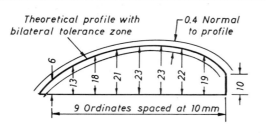
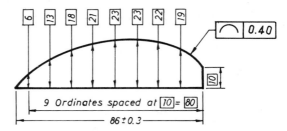

Figure 20.72 Profile of a Line

Profile of a Surface

This tolerance controls the contour of an entire surface and consequently is a three-dimensional zone across the entire area of the feature. The tolerance may be equally spaced (bilateral) or all on one side (unilateral). In the example of Figure 20.73, the tolerance zone is to be contained by upper and lower surfaces which contact the circumference of spheres 0.15 mm diameter whose centres lie on the theoretical surface form. In addition the true profile surface is to be perpendicular to datum plane A and positioned with respect to datum planes B and C.

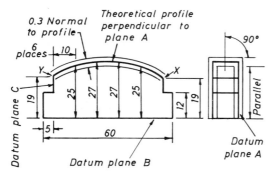
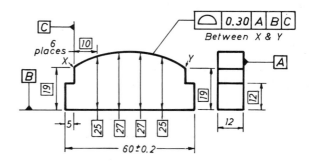

Figure 20.73 Profile of a Surface

Parallelism

By definition two parallel lines or surfaces are separated by a constant distance. In practice lines or surfaces may have to be parallel to some datum within a certain tolerance. In the example, Figure 20.74 the axis of the hole must be parallel to the end of the part. That is, the axis must be contained within two straight lines 0.05 mm apart, these lines being parallel to the end face, datum Y.

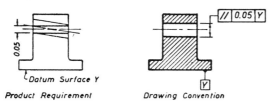

Figure 20.74 Parallelism

Squareness

This applies to the condition when a line or surface is at right angles to some datums. For example the axis of the upright part, Figure 20.75, must be contained within two straight lines 0.3 mm apart, these lines being perpendicular to the bottom datum surface. This implies that squareness is controlled in one plane only.

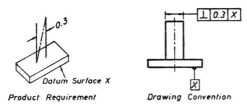

Figure 20.75 Squareness

Angularity

Angularity relates to an axis or a surface which is at some specified angle (except 90°) from a datum axis or plane.

Example 1. In Figure 20.76 the inclined surface must be contained within two parallel lines 0.3 apart and at an angle of 40° to the datum surface A.

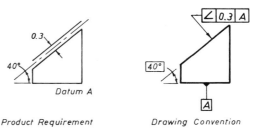

Figure 20.76 Angularity (a)

Example 2. The axis of the hole must lie within a cylindrical tolerance zone 0.2 diameter inclined at 52° to the datum plane A, Figure 20.77.

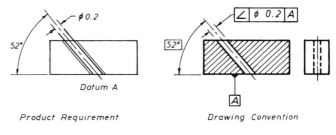

Figure 20.77 Angularity (b)

Runout

This is a geometrical tolerance requiring a practical test where the part is rotated 360° around its axis. The results of the test may include other errors such as errors in roundness, concentricity, perpendicularity, or flatness and the simple test cannot discriminate between them. The tolerance is equal to the full indicator movement of a fixed point measured during one revolution of the part about its axis.

Example 1. Runout must not exceed 0.3 at any point along the cylinder measured perpendicular to the datum axis, Figure 20.78.

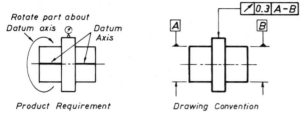

Figure 20.78 *Runout (a)*

Example 2. Runout must not exceed 0.2 at any point along the cylinder measured perpendicular to the datum diameters, Figure 20.79.

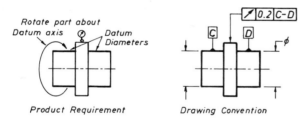

Figure 20.79 *Runout (b)*

Position

Tolerance of position controls the location of one feature from another feature or from a datum. The tolerance zone itself may be the space between two parallel lines or planes, a circle or a cylindrical space. In the example of Figure 20.80 the axis of the hole must be contained in a cylindrical zone 0.2 diameter and its axis must be coincident with the true position of the hole axis.

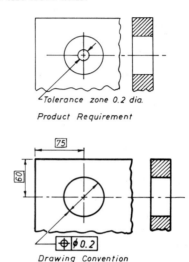

Figure 20.80 *Position*

Concentricity

Concentric cylinders are ones which have the same axis. In the example of Figure 20.81 the right-hand cylinder must be contained within a cylindrical tolerance zone which is concentric with the axis of the left-hand datum cylinder.

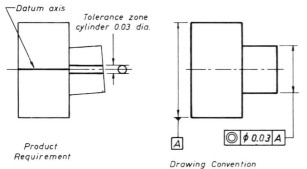

Figure 20.81 Concentricity

Symmetry

This concerns the spacing of features of a part so that they are equally positioned relative to a datum line or plane. The tolerance zone is the space between two parallel planes or lines, parallel to and placed symmetrically about the datum. In the example of Figure 20.82 the median plane of the slot and tongue must lie between two parallel planes 0.3 mm apart.

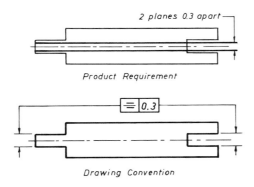

Figure 20.82 Symmetry

Maximum Material Condition—Applications

As previously stated, the maximum material condition (MMC) having the symbol Ⓜ means that a feature contains the maximum amount of material permitted by the tolerance.

When two parts mate to form an assembly then any errors of form or position have the effect of changing the respective sizes of the parts. The tightest assembly occurs when each feature is at its MMC, plus the maximum errors permitted by any geometrical tolerance.

The assembly clearance will be increased (a loose fit) if the actual sizes of the mating features are not at their MMC, and any errors of form or position are less than that specified by any geometrical control. In addition, if either part is finished away from its MMC, the clearance gained could permit an increased error of form or position to be accepted. An increase of tolerance obtained this way, provided it is acceptable functionally, is advantageous during the manufacturing process.

The concept of MMC can be applied to the characteristics of straightness, parallelism, squareness, angularity, position, concentricity, and symmetry. It cannot be applied to flatness, roundness, cylindricity, profile of a line or surface, or runout.

Maximum Material Condition Applied to Straightness

As an example, consider the pin shown in Figure 20.83(a). There are limits on the diameter and also a straightness tolerance of 0.3 applicable at the MMC.

Suppose now that the pin is made to the MMC and the maximum straightness error exists. The effective diameter if the pin is to be assembled in a hole, is equal to the sum of the upper limit of size plus the straightness tolerance, Figure 20.83(b).

If the pin should be finished to its low limit of size and we wish to retain the same assembly diameter of 16.7 then a straightness error as large as 0.7 could be acceptable, Figure 20.83(c).

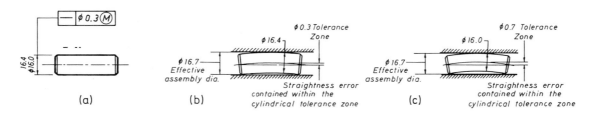

Figure 20.83 Maximum Material Condition applied to Straightness

Maximum material condition has similar applications in regard to position, concentricity, squareness, etc. Examples will be found in the text listed in the References.

Tolerances on Position or Location—Cumulative Tolerances

As an illustration the holes in the part shown in Figure 20.84(a) are located by toleranced dimensions. The hole centres between each pair of holes are tolerances separately, therefore the limits are cumulative. Between holes A and B they are ±0.1, between A and C they are ±0.2, and between A and D they are ±0.3.

An alternative method is one in which each hole is located from a common datum which in Figure 20.84(b) is the centre line of hole A. The distance between holes B and C could be 20.2 or 19.8 depending on whether B or C are positioned at their high limits or low limits. Similarly the distance between holes B and D could be 40.2 or 39.8, i.e., 40 ± 0.2. It can be seen that using this method no further accumulation of error results after the first pair of hole pitches when used on a long series of holes. However, if the acceptable error is only ±0.1 then the drawing must have limits of ±0.05 applied.

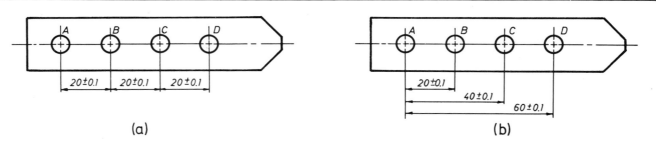

Figure 20.84 Tolerances on Position (1)

True Position (Basic) Dimensioning

The exact location of a feature is defined by true position dimensioning. These dimensions are always shown boxed on a drawing. They are never individually toleranced and therefore must be accompanied by a positional tolerance for the feature to which they are applied. The North American term for a true position dimension is a "Basic" dimension.

The main advantages are:

1) there are no cumulative tolerances;
2) the boxed dimensions fix the exact position of a feature, therefore interpretation of the drawing is easier;
3) interchangeability can be ensured without using small position tolerances as required by coordinate tolerancing systems;
4) the tolerancing of complicated parts is made simpler;
5) the positional tolerance zones can control squareness and parallelism.

Example 1. The axes of the four holes must be contained within cylindrical tolerance zones 0.03 mm diameter, Figure 20.85(a).

Example 2. The axes of the four holes must be contained within rectangular tolerance zones 0.04 x 0.02, Figure 20.85(b).

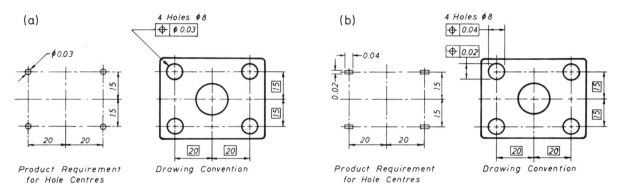

Figure 20.85 *Tolerances on Position (2)*

Example 3. In the example of Figure 20.86(a) control is required of the two φ12 holes with respect to the datum hole A. The drawing convention implies that the two φ12 holes must be in line, because of the boxed dimension 25, but subject to a positional tolerance of φ0.03. In addition their centres are subject to variation by virtue of the positional tolerance given to the φ15 datum hole. An extreme but acceptable condition for the position of the hole centres is shown in Figure 20.86(b).

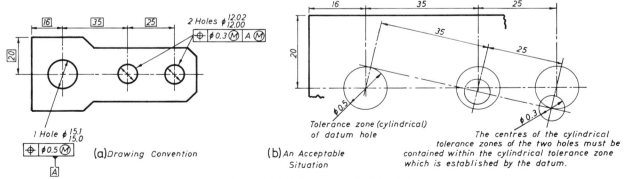

Figure 20.86 *Tolerances on Position (3)*

PROBLEMS

P 20.1 The front and side view and a sectional view of an object are shown in Figure P 20.1. Draw a freehand sketch of the top view in the correct position. It is suggested that the drawing be made on a sheet of tracing paper temporarily taped over the page. Draw sections A–A and B–B freehand and locate them adjacent to the side and front views.

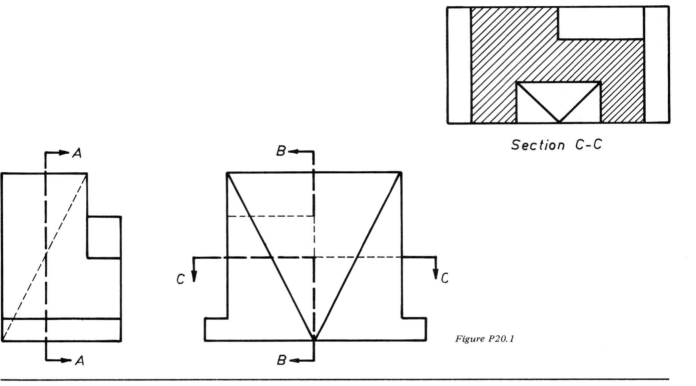

Figure P20.1

P 20.2 Third-angle projections of an object are shown in Figure P 20.2. Make an isometric sketch and answer the following questions.

a) Is surface G in front of or behind surface H?
b) Which of the surfaces G,H,C,K are in the same plane?
c) Which is the nearer surface, C or K?
d) Is surface F above or below surface A?
e) Which two surfaces are the highest?
f) Are surfaces B and D on the same level?
g) Which surface is indicated by the hidden detail on the top view?
h) Which surface is indicated by the hidden detail on the front view?
i) Which surface is indicated by the hidden detail on the side view?
j) Which of the surfaces M,N,R,P are in the same plane?
k) Which surface is on the extreme left?

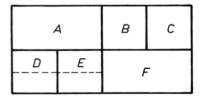
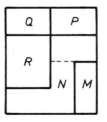
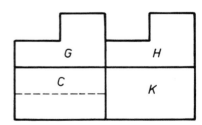

Figure P20.2

P 20.3 Draw a freehand sketch, in the correct position, of the auxiliary view of the part shown. Draw on tracing paper taped over Figure P 20.3.

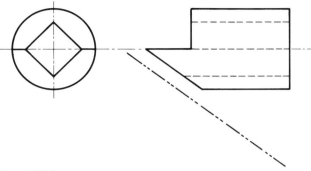

Figure P20.3

P 20.4 Draw a freehand sketch showing sections A–A and B–B. Sketch on tracing paper taped over Figure P 20.4. Also sketch a right-hand side view.

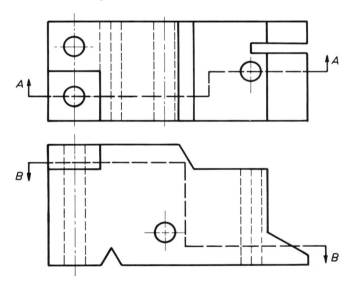

Figure P20.4

P 20.5 A section of piping is shown in an isometric sketch. Orthographic projections, in pairs, are shown for a number of piping assemblies. Which assemblies are also shown in the isometric sketch?

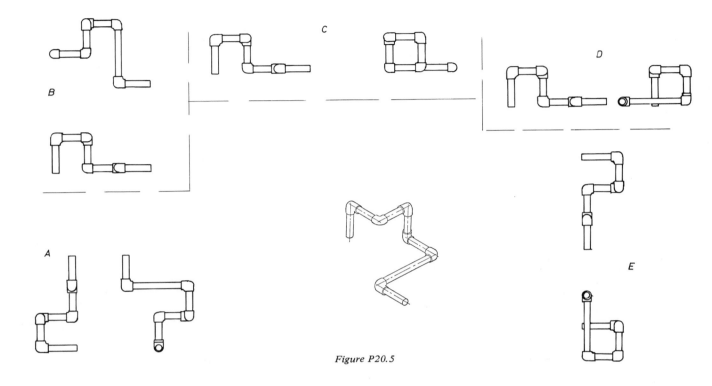

Figure P20.5

P 20.6 A part is shown in third-angle projection in Figure P 20.6.

a) Is surface A in front of or behind surface B? (b) Determine dimension E.
b) Is surface C in front of or behind surface D? What is the distance between surface C and D?
c) Determine dimension F. (d) Find dimension G. (e) Find dimension H. (f) Find distance J.
g) How far is surface K below the top face? (h) What is distance between surfaces L and M?
i) Is surface N above or below surface P? What is the distance between them?

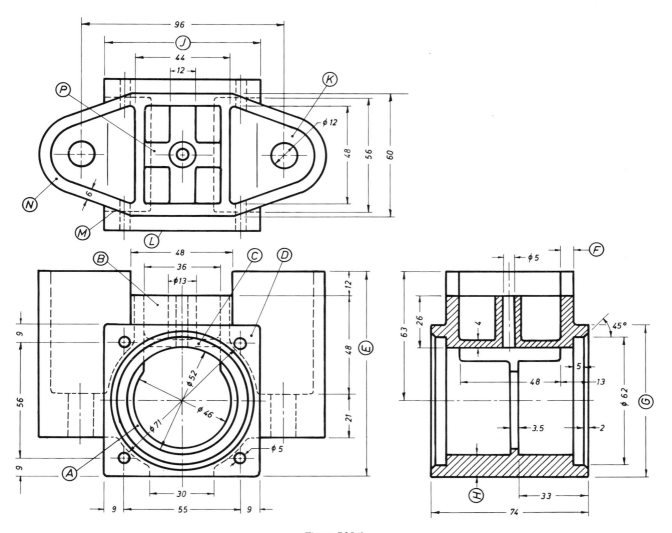

Figure P20.6

P 20.7 The drawing shows a partly dimensioned gearbox cover. Many dimensions have been omitted to emphasize the boxed dimensions. Answer the following questions.

a) How many datums are there? Give a logical reason for each one.
b) If the shaft to fit in hole E is dimensioned $54 \begin{subarray}{l} +0.000 \\ -0.010 \end{subarray}$ what is the maximum clearance? What is the minimum clearance?
c) Why is surface A chosen as a datum; why not the upper surface of the cover?
d) What range of surface finish values would be suitable for the bores of the holes?
e) What method or methods of machining could produce the surface finish values necessary for tolerances?

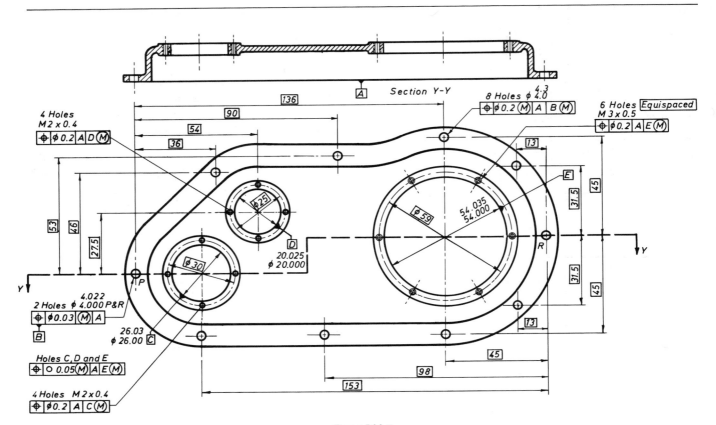

Figure P20.7

P 20.8 The cylinders A and B are each required to have their axes straight within a cylindrical tolerance zone of 0.03 mm, Figure P 20.8. Show this requirement in the correct geometrical tolerance frame attached to the drawing.

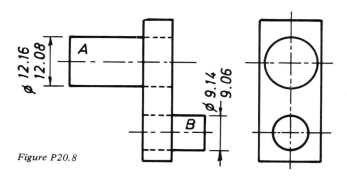

Figure P20.8

P 20.9 The surfaces C and D must be flat within a tolerance zone of 0.05 and 0.06 respectively. Show this by geometrical tolerancing on Figure P 20.9.

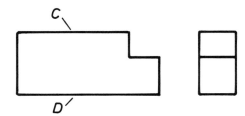

Figure P20.9

P 20.10 It is required that the part shown in Figure P 20.10 be round within 0.05 mm. Indicate this by geometrical tolerancing.

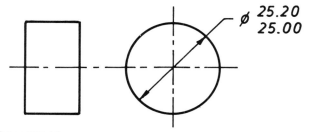

Figure P20.10

P 20.11 The cylinder shown in Figure P 20.11 has a cylindricity requirement that the surface must lie within an annular tolerance zone 0.06 wide. Indicate this requirement using geometrical tolerancing on the drawing.

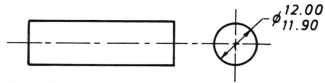

Figure P20.11

P 20.12 The bar in Figure P 20.12 must have the top surface A parallel to the bottom surface B within a tolerance of 0.07. Indicate this by geometrical tolerancing.

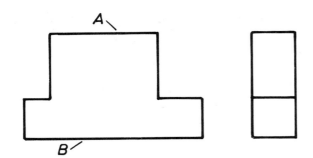

Figure P20.12

P 20.13 The part shown in Figure P 20.13 must have the side C perpendicular to the base D within a tolerance zone 0.10 wide. Show this requirement by geometrical tolerancing on the drawing.

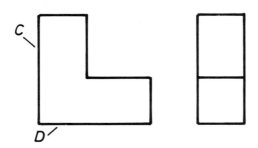

Figure P20.13

P 20.14 The hole in the part shown in Figure P 20.14 is required to be at an angle of 60° to the top surface. Indicate this by geometrical tolerancing on the drawing showing the tolerance zone to be a cylinder 0.3 diameter.

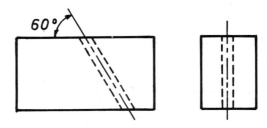

Figure P20.14

P 20.15 The part shown consists of two cylinders which must be concentric. Show the method of specifying this on the drawing by using geometrical tolerancing symbols. The tolerance is 0.04 and the datum is to be the cylinder on the right.

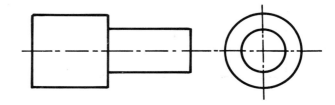

Figure P20.15

P 20.16 Show the true position of the holes in the part in Figure P 20.16 by geometrical tolerancing. Use a tolerance zone 0.04 diameter.

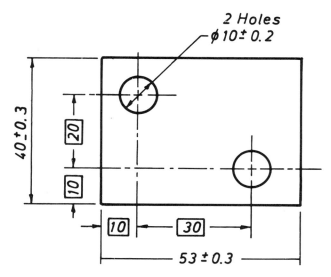

Figure P20.16

P 20.17 The part shown is to be symmetrical about its centre line. Indicate this by geometrical tolerancing using a tolerance zone of 0.2.

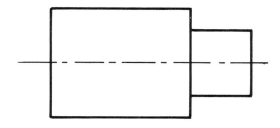

Figure P20.17

P 20.18 The part shown in Figure P 20.18 must be made so that the circular runout of surface A is within a tolerance zone of 0.05 with respect to the axis. Show this requirement by geometrical tolerancing on the drawing.

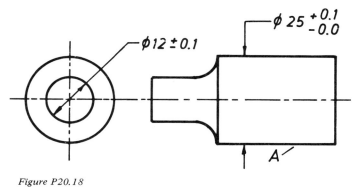

Figure P20.18

21. Freehand Technical Sketching

Freehand drawing, that is drawing without the aid of instruments, is an essential part of engineering drawing. The ability to produce a neat sketch which can be clearly understood is a valuable tool to all those concerned with communication of technical information and ideas. In addition freehand sketching is an aid to creativity in developing ideas. It is an aid to clear thinking and the logical development of new designs.

Artistic ability is not essential in making freehand technical sketches. What is required is the ability to draw straight and curved lines, accurate observation and thinking, and a sense of proportion. All these can be acquired by practice by those with average intelligence. Freehand sketches must not be careless or untidy, but should be reasonably accurate drawings.

Freehand technical sketching involves the freehand drawing of the standard orthographic views (top, front, side, etc.) and the drawing of pictorial views (oblique, isometric, perspective, etc.); see Figure 21.1. That is, we must have a sound knowledge of the above concepts as the sketches are based on those principles. For this reason this chapter has been put at the end of the text. However, practice in freehand sketching should be carried out right from the start of the study of engineering drawing. It is suggested that before any drawing exercise or problem is started with instruments it should first be sketched quickly and accurately. The preliminary sketching will force one to think first about the layout and what one intends to do before rushing to the drawing board. A preliminary sketch will clarify thinking and usually result in avoiding errors and saving time.

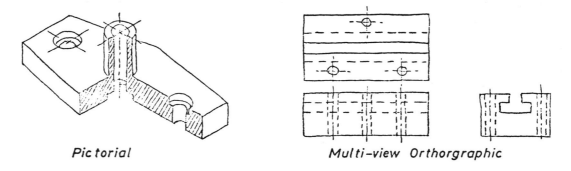

Figure 21.1 Types of Freehand Sketches

21.1 EQUIPMENT

The minimum equipment required for sketching is, of course, a pencil and some paper. The pencil usually preferred is soft (3B to HB) because of the blacker lines produced and the need to erase frequently as ideas grow or changes suggest themselves. Plain white paper is generally used, some prefer paper with faintly ruled squares to assist in proportioning. There are also available specially ruled papers with an isometric grid or a perspective grid printed as guidelines, Figure 21.2. An isometric or perspective grid can be carefully drawn and used as an underlay for plain paper.

21.2 PENCIL AND LINE TECHNIQUES

Hold the pencil just firmly enough to control the movement, resting the knuckle of the little finger on the paper. The paper should not be taped to the drawing surface, as it is convenient to move it about to suit individual preferences.

Long straight lines are quite difficult to draw freehand. This is because the hand and wrist are so constructed that short movements are quite precise while long movements are not so easily made. Long lines may be constructed using a motion of the whole arm, or they may be broken down into a series of short strokes. When drawing a line from A to B (Figure 21.3) it may be helpful to locate B by a dot, start at A with the pencil, fix the eye on B and draw from A to B. Draw lightly and erase and correct if necessary. If satisfactory, then one can trace over the line with a heavy stroke.

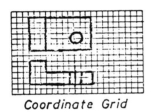
Coordinate Grid

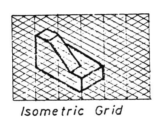
Isometric Grid

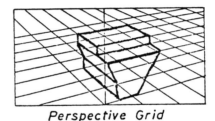
Perspective Grid

Figure 21.2 Sketches Made on Prepared Grids

Figure 21.3 Straight Lines

Vertical downward strokes can be controlled much easier than vertical upward strokes. Unless they are very short, upward strokes should be avoided. Lines which slope upward from left to right may be drawn either from bottom to top or vice versa.

Small circles up to about 50 mm diameter may be sketched by first drawing a circumscribed square. Lightly sketch the centre lines. Then lightly outline each quadrant, making sure the line is tangent to the sides of the square; see Figure 21.4(a). After sketching lightly, correct any deviations by erasing, then finally go over the complete circle with a firm heavy line. Alternatively, and for larger circles, additional radial lines may be drawn lightly and the radial distance marked off by eye or by using a piece of paper as a length guide, Figure 21.4(b). Arcs are then sketched in perpendicular to each radial line.

Ellipses occur frequently in pictorial drawings. As an aid, the enclosing rectangle may be drawn first, with its centre lines. Lightly draw in the tangents and then join the tangents with curved lines. Correct and erase as necessary and then firm in with a heavy outline, Figure 21.5. Note that the ellipse is almost tangent to the line AB. A and B are each located 1/4 of the distance along the edges of the rectangle.

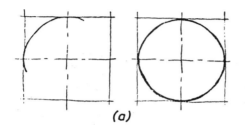
(a)
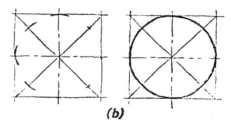
(b)

Figure 21.4 Circles

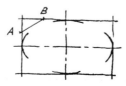
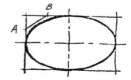

Figure 21.5 Ellipses

21.3 CRATING

As an aid to pictorial sketching, most objects can be divided up into crates or boxes which may surround rectangular parts, cylinders, cones, spheres, etc. These crates can be sketched lightly in the required position to represent the object. The crates are then filled in by sketching the particular form they surround. As an example, the bearing block shown in Figure 21.6(a) may be imagined to consist of the crates or boxes shown in Figure 21.6(b).

The centre lines of the crates are sketched in as an aid in drawing the ellipses. Note that the complete ellipse should be drawn at first, including the hidden parts. The final erasing and firm outlining will result in the finished sketch as in Figure 21.6(a). Dimensions can be added if necessary; see Section 12.8.

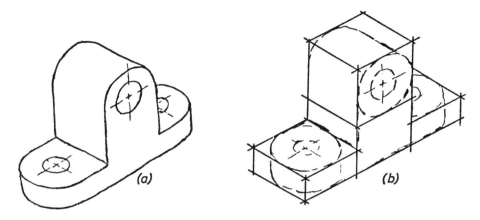

Figure 21.6 Crating or Boxing

21.4 PROPORTION

Correct proportion is combined with crating in freehand sketching. The essential proportions of an object must be realized when constructing the crates. This is similar to the work of a sculptor who starts with a block and then removes unwanted material to produce the final result. As an example consider the orthographic views of the bracket in Figure 21.7(a). We may first draw a large crate to contain the whole object, Figure 21.7(b). Noting that the central part is about twice as large as each end piece we can divide this crate into three parts, Figure 21.7(c). The lower parts are about one-sixth the height of the crate so that they can now be roughed in. Locate centre lines of the holes and box them in, Figure 21.7(d). The completed sketch is shown in Figure 21.7(e).

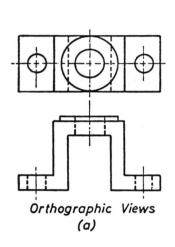
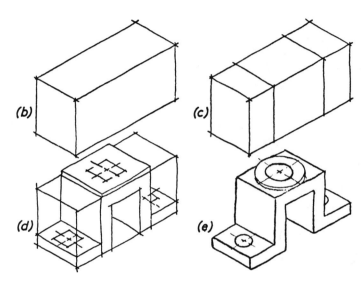

Figure 21.7 Proportioning

21.5 PICTORIAL AND ORTHOGRAPHIC SKETCHING

Pictorial sketches are a considerable aid to visualisation of an object. Thus they are much preferred over the orthographic. Orthographic sketching is useful, but the views are flat projections, showing top, front, side, etc., views. Orthographic sketching is the freehand application of orthographic projection principles and drawing practices.

The most common type of pictorial sketches are oblique, isometric, and perspective, the basic theory of these being dealt with in Chapter 12. Other useful types are dimetric and trimetric.

21.6 OBLIQUE SKETCHING

The advantage of this method is that when the front face is drawn parallel to the picture plane it appears in true shape. Therefore, sketch the object with the important or irregular detail in true shape. The disadvantage is that the resulting sketch may appear distorted. As indicated in Section 12.3, the foreshortening is one-half for Cabinet drawing and 1.0 for Cavalier drawing. An example is shown in Figure 21.8. The orthographic views are shown in Figure 21.8(a). First lay off the axis XY at any convenient angle. We do not need to sketch in crates or boxes but merely lay off the centres of the circles along the axes, Figure 21.8(b). Draw enclosing squares, Figure 21.8(c). Sketch in circles in true shape and complete. A comparison is shown between Cabinet and Cavalier in Figs. 21.5(d) and (e).

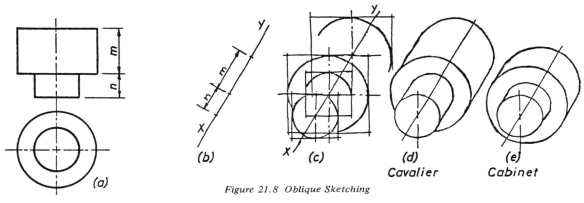

Figure 21.8 Oblique Sketching

21.7 ISOMETRIC SKETCHING

Isometric drawing is described in Section 12.6. Isometric freehand sketching is the freehand application of those principles. The advantage of isometric is that it is simple and quick. A disadvantage is that it produces some distortion. Circles in isometric become ellipses as explained in Section 12.7. An isometric projection has equal foreshortening along all three isometric axes, but an isometric drawing or sketch is produced using true lengths along the isometric axes. Any line not parallel to an isometric axis does not appear in true length and must be drawn by plotting contact points, Figure 21.9. Curves may be sketched by similarly plotting a number of points on the curve and joining the points.

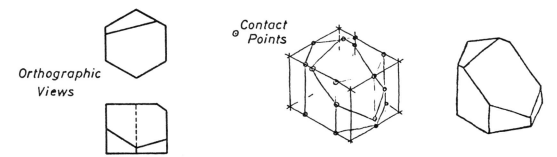

Figure 21.9 Isometric Sketching

The isometric axes are 120° apart but they may be inverted as shown in Section 12.6. They may also be rotated to any position required to produce a more suitable view. Some examples are shown in Figure 21.10.

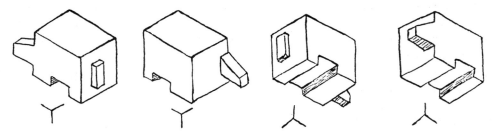

Figure 21.10 Rotation of Isometric Axes

21.8 DIMETRIC AND TRIMETRIC SKETCHING

In dimetric sketching two of the principal axes are equally foreshortened, and in trimetric sketching none of the axes are equally foreshortened. ("Iso" means "all" are equally foreshortened; "di" means "two" are equal; and "tri" means "three" are unequal.) While dimetric and trimetric sketches may sometimes be more realistic appearing than isometric, measuring and plotting are more difficult. Some typical positions for dimetric and for trimetric drawing are shown in Figure 21.11. The ratios of lengths along the axes are shown along the edges of each cube. In freehand sketching the angles and ratios are, of course, estimated by eye.

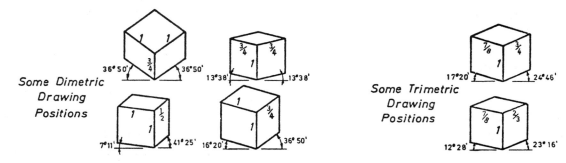

Figure 21.11 Dimetric and Trimetric Drawings

21.9 PERSPECTIVE SKETCHING

The principles of perspective projection are detailed in Chapter 12. Perspective sketching is the freehand application of those principles. The advantage is that the resulting sketch is more life-like than other pictorial methods. The disadvantage is the difficulty of measuring and plotting curved lines and circles. We will consider the sketching of a cube in two-point perspective. First sketch the horizon; see Figure 21.12. At each end locate the right and left vanishing points, VPR and VPL. Sketch in a vertical line offset from the middle (about 30° and 60° at the station point is convenient. Sketch the front edge of the cube, AE. Draw lines from A and E to the vanishing points. Approximate EF and EH remembering that EF will be shorter than EH and both will be less than AE. Draw verticals from F and H and complete the sketch as shown. To avoid distortion angle FEH should be greater than 90°.

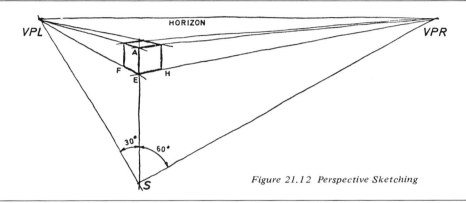

Figure 21.12 Perspective Sketching

Large Perspective Sketches

When making a large sketch in perspective, the cube may be drawn directly without actually putting in the horizon and vanishing points, Figure 21.13. However, the position of the vanishing points must be always kept in mind even though they may be off the edge of the paper. It is helpful to extend the horizontal edges of the cube a short distance towards the imagined vanishing points. Remember that vertical lines actually appear vertical in two-point perspective. Another aid is to sketch lightly the hidden rear vertical edge to ensure that it passes through the intersection of the rear bottom edges as at G.

In the case of tall objects, such as buildings, the horizon may be below the top surface so that the top edges will slope down as indicated in Figure 21.14.

Figure 21.13 Large Cube

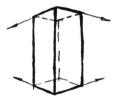

Figure 21.14 Tall Object

Repetition

When dividing an object up into crates or boxes the situation often arises that the crates are repeated. Little difficulty occurs in duplicating a box in isometric sketching, however in perspective it is not so easy. One method is to sketch a line from one corner of a square A, through the mid-point of the opposite side, B, to find the correct position for the next square, F, Figure 21.15. The process may be repeated as shown.

Cubes may be built up one upon the other by extending the diagonal AE to meet the vertical at D, thus developing a second layer, Figure 21.16.

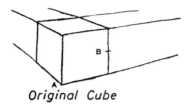
Original Cube

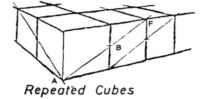
Repeated Cubes

Figure 21.15 Repetition

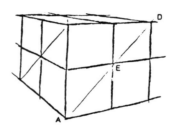

Figure 21.16 Repeated Cubes

Example 21.1

As an example of the methods consider the steps taken to produce a perspective sketch of the object shown in the orthographic views of Figure 21.17.

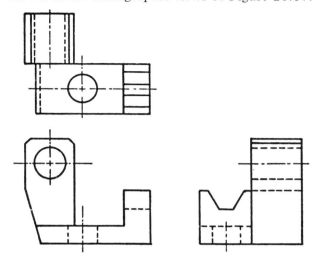

Figure 21.17 Orthographic Views

In order to divide the object into crates we can consider it to be composed of parts of five cubes as shown in Figure 21.18(a). The cubes can then be cut away starting with the right-hand end cube 5 which is only about half as wide as the other cubes. In addition we can cut cubes 3 and 4 down to about one-third their height. Also, at this stage, shave off the face of cube 2; see Figure 21.18(b). Next locate the centres of the circles and sketch in the enclosing squares. Draw the chamfer on the top of cube 1 and also sketch in the Vee groove in block 5 by visual proportioning, Figure 21.18(c). Complete by firming in the outlines and erasing the light construction lines of the original crates, Figure 21.18(d).

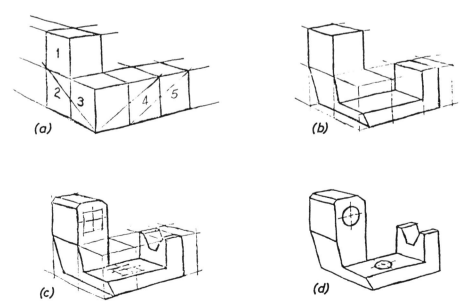

Figure 21.18 Perspective Sketching

21.10 EXPLODED PICTORIALS

Exploded views of aircraft engines, machines, etc. are very useful in visualising the assembly and the relative position of the various parts. The separation of the individual parts should be done in the direction of an axis of the whole drawing. An example of an exploded isometric sketch is shown in Figure 21.19. Note that each part is visible and that the parts have been separated along directions parallel to the isometric axes.

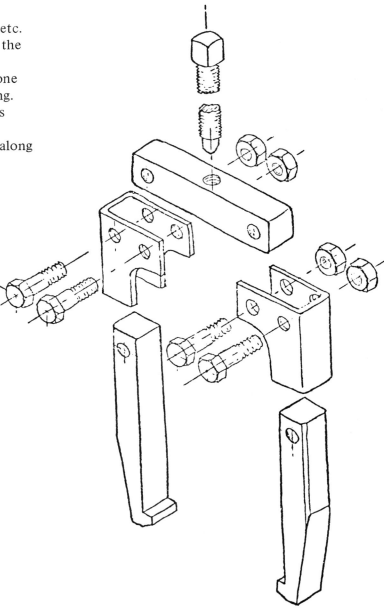

Figure 21.19 Exploded Pictorial

21.11 SHADING

Shading sometimes aids the visualisation of certain planes and curved surfaces. The purpose is to make interpretation of the sketch easier and not to ornament it. A little shading is usually more effective than too much. The pencil may be held normally or lowered to make a small angle with the paper.

Shading on a cylinder is shown in Figure 21.20(a). Try for an even graduation between light and dark areas. Various shadings on a block are shown in Figure 21.20(b). Line shading may be done more quickly, Figure 21.20(c).

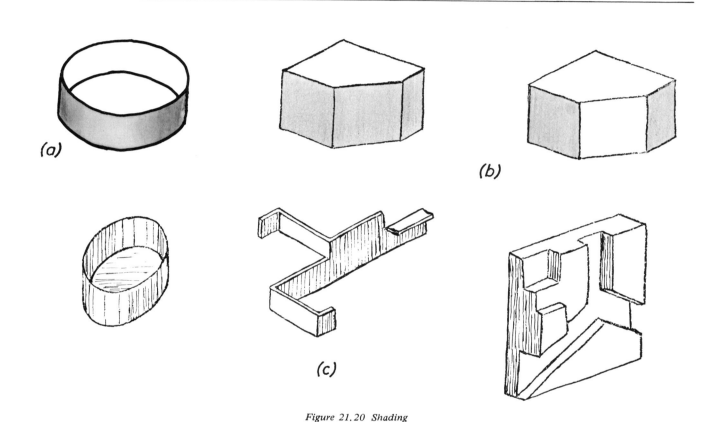

Figure 21.20 Shading

PROBLEMS

P 21.1 Sketch the design shown, first within a square of overall dimensions about 50 mm by 50 mm, then in a square about 100 mm by 100 mm.

P 21.2 Sketch the truss shown outlined in Figure P 21.2. Make the length of the truss about 100 mm, and proportion the members in the ratio of the numbers shown.

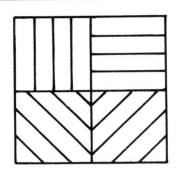

Figure P21.1

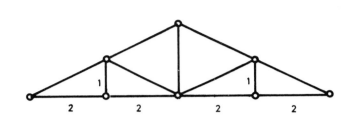

Figure P21.2

P 21.3 Sketch a series of circles, each about 25 mm diameter. Space the circles as equally as possible by eye. Figure P 21.3.

P. 21.4 Repeat P 21.3 but draw circles about 50 mm diameter.

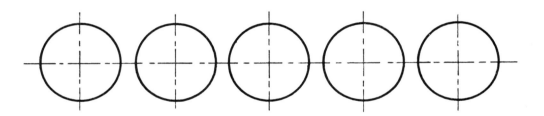

Figure P21.3
Figure P21.4

P 21.5 Sketch the series of ellipses shown. First sketch the circumscribing rectangles to the proportions indicated by the numbers. Then construct the ellipses freehand. Repeat with larger rectangles.

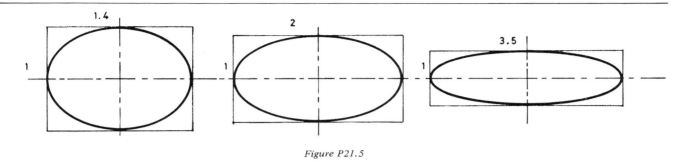

Figure P21.5

P 21.6 Make a freehand orthographic sketch of the part shown in Figure P 21.6. Make dimension X about 100 mm long. Numbers refer to ratios of lengths.

P 21.7 Make a freehand orthographic sketch of the object shown in Figure P 21.7. Make dimension Y about 100 mm long. Numbers refer to ratios of lengths.

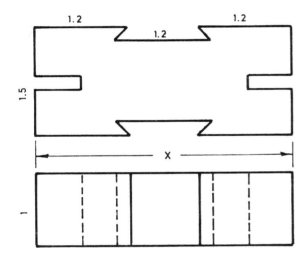

Figure P21.6

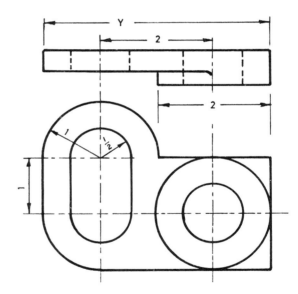

Figure P21.7

P 21.8 Sketch a series of isometric cubes as indicated in Figure P 21.8. Use a length of side of about 25 mm. **Repeat** about twice as large.

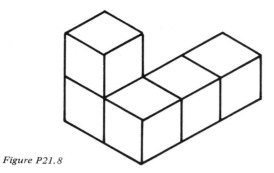

Figure P21.8

P 21.9 Draw isometric cubes of various sizes. Fit circles as in Figure P 21.9.

Figure P21.9

P 21.10 Make an isometric sketch of the object shown. Proportion it so that the longest edge is about 100 mm long.

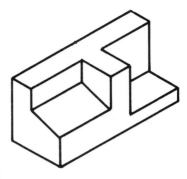

Figure P21.10

P 21.11 Make an isometric sketch of the part shown. Length of base about 80 mm.

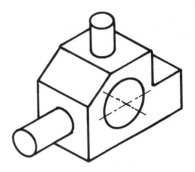

Figure P21.11

P 21.12 Make an oblique sketch of the part shown in Figure P 21.12. Length about 100 mm.

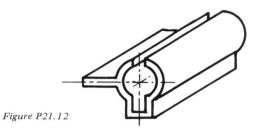

Figure P21.12

P 21.13 Make a perspective sketch of the object shown in Figure P 21.13. Total length about 100 mm.

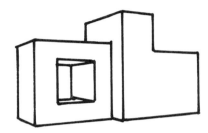

Figure P21.13

P 21.14 Make an exploded isometric sketch of the gear puller, Figure 21.19. Overall height about 200 mm.

P 21.15 Make a perspective sketch of the object shown. Take proportions from the orthographic views of Figure P 21.15(a). Depending on your choice of the position of the point of sight your sketch may not appear similar to that of Figure P 21.15(b). Use the crating method as a preliminary step.

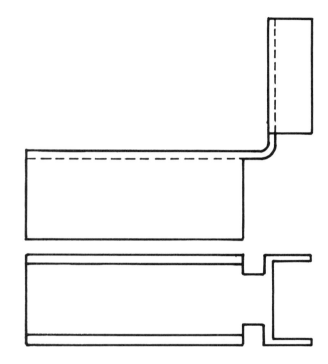

Figure P21.15(a)

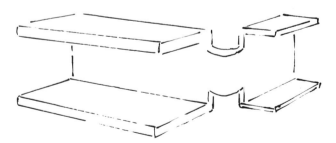

Figure P21.15(b)

APPENDIX A
Equipment

Descriptive Geometry involves the use of graphical methods to solve three-dimensional spatial problems. In order to produce accurate results it is necessary to have good quality equipment including drawing instruments. This section indicates the minimum equipment needed to produce a graphical solution to many problems.

A.1 GENERAL
It is essential that proper equipment be used with skill in order that the solution of the problems will have an acceptable accuracy. To attain this end result, straight lines must be truly straight, angles must be true, and construction lines must be fine and well defined.

A.2 PENCILS
Lead for the wooden or mechanical lead holder type of pencil is available in a number of grades designated by letters. The softest lead is grade 6B. The grades 5B, 4B, etc. are less soft by steps up to B, HB, and F. The grades then continue in increasing hardness from H up to 9H. Most of the drafting work is done using the grades from 2H to 5H and a sharp pencil.

A.3 PAPER AND FILM
A good quality tracing paper is generally used for drawings made in pencil. Copies can be made by standard reproduction process. Plastic drafting film (polyester base) is particularly useful for ink drawings as the ink can be easily removed with a plastic eraser.

A.4 DRAWING BOARD
The drawing surface must be smooth and flat. It must have a perfectly straight guiding edge if a T-square is used. The surface is usually covered with a heavy green plastic material which may be replaced if necessary. The board may be used with a T-square or a parallel rule for drawing horizontal lines and also as a base for the drafting triangles. Some boards have a drafting head attached to the upper edge. This device combines the function of the T-square or parallel rule and the drafting triangles.

A.5 DRAFTING TRIANGLES

Triangles made of clear plastic are used in connection with the T-square or parallel rule to draw lines at various angles with the horizontal. There are two standard triangles, the 30° – 60°, and the 45° triangles. They should be handled with care. A triangle may be damaged if it is dropped or struck. The edge may become irregular or a corner raised so that it is inaccurate. Also scratched surfaces collect graphite from pencils and tend to make the drawing dirty.

To draw a line the pencil should be held at an angle of about 60° to the paper as shown in Figure A.1(a). It may be held in a plane parallel to the edge of the triangle as shown in Figure A.1(b) or in the corner as shown in Figure A.1(c). Whichever situation is chosen, the pencil should be maintained in that position while the line is being drawn.

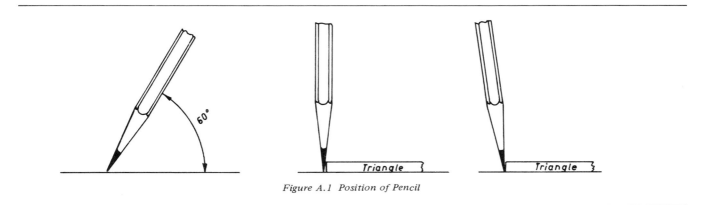

Figure A.1 Position of Pencil

A.6 TESTING OF T-SQUARE OR PARALLEL RULE

The edge of the T-square or parallel rule may not be straight. To test for straightness draw a fine line across the length of the paper. Remove the paper and rotate it 180° as shown in Figure A.2. The line and the edge of the square should coincide.

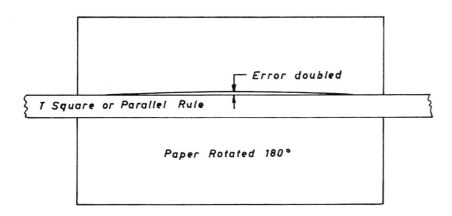

Figure A.2 Testing a T-Square or Parallel Rule

A.7 TESTING THE 90° ANGLE OF A TRIANGLE

Use a previously tested T-square or rule and draw a vertical line as shown in Figure A.3. Reverse the triangle so as to place it on the other side of the line. The edge of the triangle and the line should coincide. If not look for a damaged corner on the triangle. A raised corner can be smoothed off by using fine sandpaper.

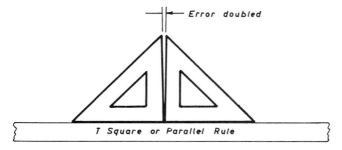

Figure A.3 Testing the 90° Angle

A.8 TESTING THE 45° ANGLE OF A TRIANGLE

Use a previously tested T-square or rule and a previously tested 90° angle. Draw a line at 45° as shown in Figure A.4. Reverse the triangle and compare the 45° edge with the drawn line. The edge and line should coincide.

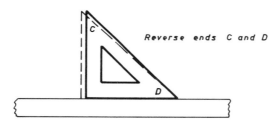

Figure A.4 Testing the 45° Angle

A.9 TESTING THE 60° ANGLE OF A TRIANGLE

Use a tested T-square or rule and draw a horizontal line AB as shown in Figure A.5. Move the T-square or rule down slightly and draw lines along the 60° edge to form an equilateral triangle as shown. Measure the three sides and if they are equal, the triangle is accurate. The 30° angle will be correct if the 90° was correct.

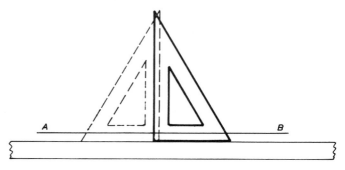

Figure A.5 Testing the 60° Angle

A.10 COMPASS

The large bow compass, Figure A.6, is used to draw circles and arcs of circles. To draw a circle first locate the centre and then mark off the radius. Place the pivot point accurately on the centre mark and adjust the position of the pencil point to the exact radius. The circle is normally drawn by tilting the instrument forward slightly and then rotating the compass in a clockwise direction. The lead may be sharpened to a chisel point which will retain its sharpness longer than a conical point.

A.11 DIVIDERS

Dividers Figure A.7, used for transferring distances on a drawing or for dividing a line into a number of equal parts. Dividers should not be used directly on a scale. Besides being inaccurate, the sharp points will damage the fine graduation marks on the scale.

A.12 PROTRACTOR

The protractor is used for drawing angles or for measuring the angle between two lines. Located accurately it should be possible to measure an angle to within less than ±½° with a 6-inch diameter protractor.

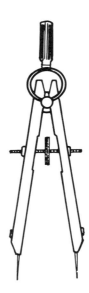

Figure A.7 Dividers

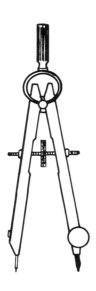

Figure A.6 Bow Compass

A.13 SCALES

Usually we represent on the drawing paper some object or situation which is larger than the paper. This means, of course, that we have to draw it at some reduced size if we want to fit it all on the drawing sheet. Less frequently the part will be drawn to an enlarged size. This is done for small parts such as some clock and instrument parts or printed circuit details. To facilitate all drawings we make use of scales which have various types of divisions marked on them. Scales should be used only for measuring distances and never as rulers for drawing lines.

There are a few standard scales in general use including the engineer's scale, the architect's scale, and the metric scale. The first two of these scales will probably be used much less in the next few years when the metric system is adopted in North America.

ENGINEER'S AND ARCHITECT'S SCALES

These scales are based on the inch and foot as the unit of measurement. The engineer's scale has the inch divided into decimal parts. An inch on the scale may be divided into 10, 20, 30, 40, 50, or 60 parts. The divisions are made over the whole length of the scale as shown in Figure A.8. This scale may be used for drawings based on the decimal system, the 50 scale being used for dimensions such as 3.62". It is also used for scaling vectors. For example the dimension concerned may be 3600 lbs. to some such scale as 1" = 3000 lbs. using the 30 scale on the rule. The engineer's scale may be used for either reduced size or enlarged size drawings.

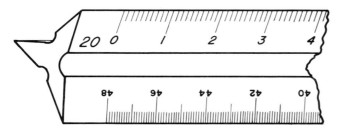

Figure A.8 Engineer's Scale

The architect's scale, Figure A.9, is based on the foot with the foot division at one end divided into inches, half-inches, or quarter-inches depending on the size of the scale. This scale may be used for drawings dimensioned in fractions of an inch, for example $3\frac{5}{8}''$. It may also be used for reduced or enlarged drawings. The usual scales being $\frac{1}{8}'' = 1$ foot, $\frac{1}{4}'' = 1$ foot, $\frac{1}{2}'' = 1$ foot, etc.

When stating a scale such as $1'' = 1$ foot or $\frac{1}{4}'' = 1$ foot or $1'' = 3000$ lbs. the first number or fraction always refers to the drawing. The second number refers to the object or the situation which is being portrayed on the paper.

METRIC SCALES

Metric scales are usually based on the centimetre or millimetre. For full-size drawings the reduction of course is 1:1 and the scale is divided into millimetres with centimetres indicated as shown in Figure A.10. For drawings to a reduced size many different scales are commercially available. For half-size drawings a 1 mm = 2 mm scale would be used, the scale being marked 1:2. Some metric scales are shown in Figure A.10.

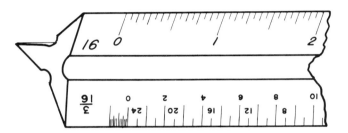

Figure A.9 Architect's Scale

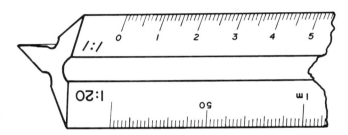

Figure A.10 Metric Scale

APPENDIX B
Accuracy

B.1 DRAWING SCALE

The accuracy which can be obtained using graphical methods obviously depends on the care with which the graphical work is performed. It also depends very greatly on the scale used. For example an error of 0.4 mm in measuring the length of a line might occur when doing careful work. If this error was made in measuring a line 190 mm long the percentage error would be only 0.2%. However if the line was only 19 mm long, the percentage error would be 2%. Therefore you should use as large a scale as your drawing board and equipment will permit.

B.2 INTERSECTING LINES

It is often necessary to locate the intersection of two lines. The point of intersection will have more chance of being in error when the angle of intersection between the two lines is small. If we can expect an error ±0.4 mm inch in the correct position of a line then the zone in which the line may lie is a thin rectangular area 0.8 mm inch thick (Figure B.1). If two lines meet at 90° then the intersection may lie anywhere within 0.8 mm square as shown in Figure B.1(a).

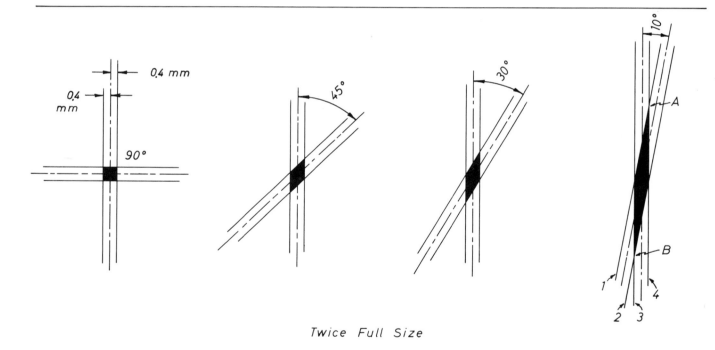

Twice Full Size

Figure B.1 Possible Errors in Intersecting Lines

As the angle between the two lines becomes less than 90°, the possible area of intersection becomes larger as shown in Figure B.1(b) and B.1(c). When the angle is 10° the possible error in locating the point of intersection is quite large. The extreme positions of the intersection will be at point A when lines 1 and 4 are used, or point B when lines 2 and 3 are used. This is about eight times as large as the range of error for the 90° lines.

In some situations the error arising from small angle intersections can be reduced by using a different graphical procedure. For example suppose that we have been given the frontal projection only of the point C on the line AB, Figure B.2. We wish to locate the horizontal projection of C. If we project upwards from C_F, the location of C_H may be in error due to the small angle of intersection. However, we can use an additional view, in this case the profile view and transfer the distance X from the profile view to the horizontal view. This latter method will be more accurate as the intersecting lines meet at large angles.

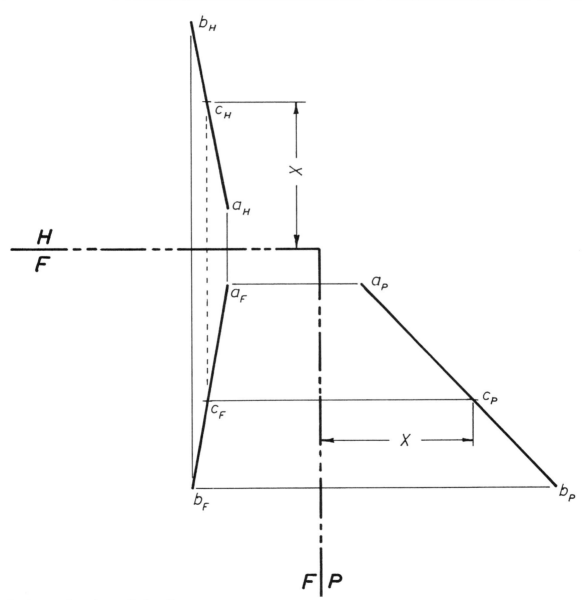

Figure B.2 Alternate Procedure to Reduce Error

For another example of a method of reducing intersecting line errors consider the location of point P in the plane ABC, Figure B.3(a). Here we are given the horizontal projection only of point P. The usual method of locating p_F would be to choose a line through p_H such as the line $d_H e_H$. Projecting down to the frontal view would locate p_F at the intersection of two lines at a small angle. The likelihood of a large error in this construction can be reduced by choosing a construction line such as $b_H f_H$ and containing p_H. The location of p_F would now be found at the intersection of two lines at a much larger angle as shown in Figure B.3(b).

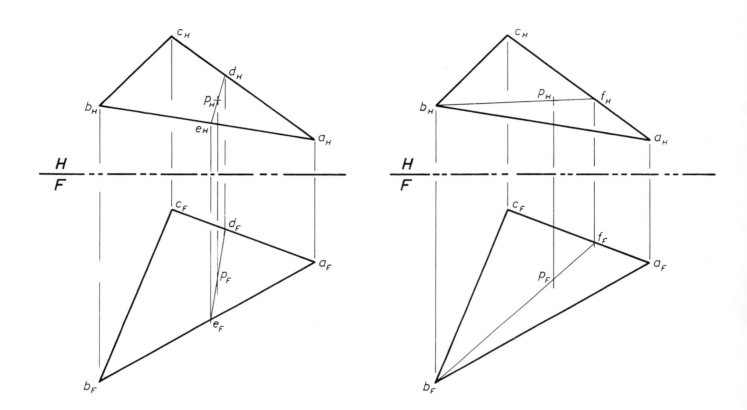

Figure B.3 Alternate Construction to Reduce Error

B.3 DIRECTION OF LINES

When a line is determined by joining two points, the direction of the line may be in error if the two points are too close together. As an example assume the range of the limit of accuracy in locating a point to be a circle 0.8 mm in diameter. Consider two points as shown in Figure B.4(a) and for simplicity assume that the line is drawn accurately through the left point. In this case the possible directions of the line may lie anywhere between the outer limit lines shown in Figure B.4. If the two points are further apart as in Figure B.4(b), then the possible error is much less for the same total projection distance.

B.4 CUMULATIVE ERROR

A small error if added in succession can amount to a significant error. This can occur when a series of equal distances are to be marked off along a line. If dividers are used, set carefully to the required short distance, any error in the original setting will be multiplied by the number of times the distance is transferred. It is better in this case to divide the line by the parallel line method. A scale may be used to divide a line into equal distances, instead of the dividers. In this case a large error can accumulate if the scale is moved between each marking off. The scale should not be shifted for this operation.

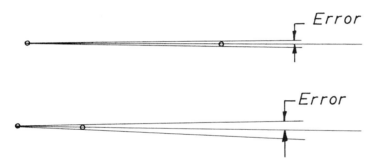

Figure B.4 Error in Direction of Lines

APPENDIX C
Lettering

365 An engineering drawing must provide more information than just the shape of the part. The dimensions must be included and also information on the material, surface finish, any special operations required, and a title. This data is lettered, usually freehand, and must be perfectly legible so that no confusion is possible. It also must be uniform. In order to accomplish this it is necessary to be able to letter rapidly and neatly. This can be done by forming the letters and numbers carefully and then obtaining uniformity and speed by practice. Lettering can improve the appearance of a drawing if it is well done, or it can ruin an otherwise excellent drawing.

SINGLE-STROKE LETTERING
This is the type of lettering used on engineering drawings. Single stroke means that the width of the stem of the letter is the width of the pencil or pen. An example of single-stroke and two-stroke lettering is shown below.

AB   *AB*

Single   Two
Stroke   Stroke

Single-stroke lettering may be either inclined or vertical. If inclined, the angle of inclination is about 68°. When using vertical lettering the vertical lines must all be vertical. Similarly when using inclined lettering the angle must be kept constant.

AB   AB

Inclined   Vertical

STYLE
The letters in the alphabet vary in width, I being the narrowest and W being the widest. Also an optical illusion exists in which a horizontal line drawn in the middle of a rectangle or rhombus appears to be below the middle. To counteract this letters such as B, E, and X must be formed smaller at the top than at the bottom. This rule also applies to the numbers 3 and 8.

UPPER CASE (CAPITALS) AND LOWER CASE LETTERS

Capital lettering is known as upper case and the small lettering is known as lower case. The terms originated when printers kept their type faces in an open wooden case, the upper portion being reserved for the capital letters. The lower portion of the case contained the small letters which were used more frequently and were therefore placed closer to hand for convenience in making up the type.

The height of lettering is referred to the height of the capitals. The stems (ascenders) of lower case letters such as b and d extend to the same height as the capitals. The body of the lower case letters are made two-thirds of the height of the capitals as indicated below. The stems of lower case letters which extend downwards (descenders) are made as far below as the ascenders are made above. These proportions are maintained by guide lines.

GUIDE LINES

Guide lines are always drawn to keep the lettering horizontal and to preserve the proportions of the letters. These may be drawn with drafting instruments or a special plastic lettering guide such as the Ames lettering instrument. In addition it is very helpful to draw lines at the correct slope for inclined lettering. A slope of 5 to 2 will provide a guide line of about 68°. The Ames lettering instrument has an edge at the required angle to facilitate the drawing of lines at the correct slope.

**FORMATION OF INDIVIDUAL LETTERS—
UPPER CASE**

The complete alphabet of single-stroke letters is shown below in both the vertical and inclined style. The order of forming each letter and the direction of the stroke is shown by small numbered arrows. A squared background is provided to clarify the proportions of widths, heights, etc.

FORMATION OF INDIVIDUAL LETTERS— LOWER CASE

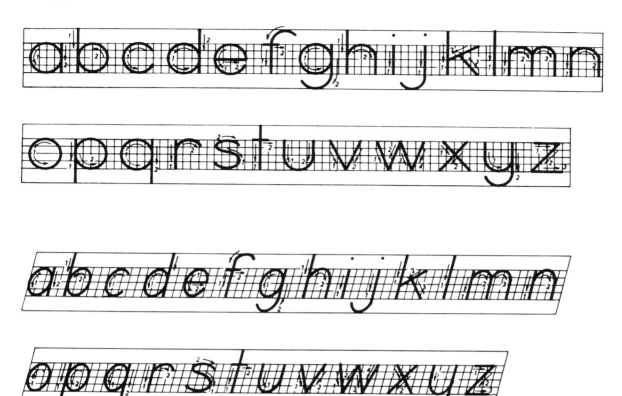

GENERAL NOTES

Letters such as O, Q, G, C, S, c, o, s should be extended slightly above and below the horizontal guide lines. If these letters are made to a height tangent to the guide lines, they will appear smaller than the other letters, due to an optical illusion. Many of the inclined letters are based on the ellipse used for o.

NUMERALS

The formation of the numerals is shown below. The height of these is made equal to that of the capitals.

For fractions the numerals in the numerator and denominator are smaller than the integral numeral. The height should be about two-thirds that of the integral. These smaller numerals should not touch the horizontal line which separates them.

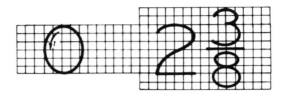

COMPOSITION

When letters are grouped to form words the spaces between the letters should be equal in area. This is not always possible but should be kept uppermost when spacing letters in a word. Also the spaces between words should be about equal to the letter o.

Bibliography

ENGINEERING DRAWING AND DESCRIPTIVE GEOMETRY

Earle, J. W. *Engineering Design Graphics.* Reading, Mass.: Addison-Wesley, 1969.

French, T. E. and C. J. Vierck. *Graphic Science and Design.* New York: McGraw-Hill, 1970.

Giesecke, F. E. and others. *Technical Drawing.* New York: Macmillan, 1969.

Grant, H. E. *Practical Descriptive Geometry.* New York: McGraw-Hill, 1965.

Hoelscher, R. P. and C. H. Springer. *Engineering Drawing and Geometry.* New York: Wiley, 1965.

Hood, G. J., A. S. Palmerlee, and C. J. Baer. *Geometry of Engineering Drawing.* New York: McGraw-Hill, 1969.

Luzadder, W. J. *Graphics for Engineers.* Englewood Cliffs, N.J.: Prentice-Hall, 1968.

Paré, E. G., R. O. Loving, and I. L. Hill. *Descriptive Geometry.* New York: Macmillan, 1971.

Shearman, R. L. *Perspective Drawing.* London: Longmans Green, 1969.

Slaby, S. M. *Fundamentals of Three-Dimensional Descriptive Geometry.* New York: Harcourt Brace, 1966.

Thomson, R. *Exercises in Graphic Communication.* London: Thomas Nelson and Sons, 1971.

Wellman, R. L. *Technical Descriptive Geometry.* New York: McGraw-Hill, 1957.

FREEHAND SKETCHING

Cook, H. *Freehand Technical Sketching.* London: Methuen and Co., 1965.

Katz, H. H. *Technical Sketching and Visualisation for Engineers.* New York: Macmillan, 1959.

Knowlton, K. W., R. A. Beauchemin, and P. J. Quinn. *Technical Freehand Drawing and Sketching.* New York: McGraw-Hill, 1977.

KINEMATICS

Tao, D. C. *Fundamentals of Applied Kinematics.* Reading, Mass.: Addison Wesley, 1967.

Shigley, J. E. *Kinematic Analysis of Mechanisms.* New York: McGraw-Hill, 1969.

PHOTOGRAMMETRY

Cruse, J. and A. Newman. *Photographic Techniques in Scientific Research.* London: Academic Press, n.d.

Hallert, B. *Photogrammetry.* New York: McGraw-Hill, 1960.

Williams, J. C. C. *Simple Photogrammetry.* London: Academic Press, 1969.

GEOMETRICAL TOLERANCING

Foster, L. W. *Geo-metrics. The Metric Application of Geometric Tolerancing.* Reading, Mass.: Addison-Wesley, 1974.

Simmons, C. H. and D. E. Maguire. *A Manual of Engineering Drawing Practice.* London: English Universities Press, 1974.

Index

Accuracy, 361–64
Alphabet, single-stroke lettering, 367
Analglyph drawings, i–iii
Angle,
    between a line and the principal planes, 115
    between a line and an oblique plane, 75, 117
    between two skew lines, 95
    specified, for a line, 119
Architect's scale, 359, 360
Architectural symbols, 316
Assembly drawings, exploded, 317
Auxiliary planes, 49
    double or inclined auxiliary, 53
Axes,
    choice of, 266
    general, 37
    isometric, for computation, 257–61
    perspective, for computation, 267–71

Basic dimensioning, 330
Basic hole system, 319
Basic shaft system, 320
Beams, forces on, 73
Bearing of a line, 49
Bibliography, 370
Bilateral limits, 319
Board, drawing, 355
Breaks, conventional lines, 302
Broken-out sections, 304

Cabinet projection, 133
Calculators, 257
Calculus,
    graphical differentiation, 227
    graphical integration, 233
Capital lettering, 367
Cavalier projection, 133
Centre lines, 302
Chordal method, 231, 234
Clearance fits, 318
Compasses, 358
Components of a force, 164
Computation table, 260

Computers, 257
Computer program,
    isometric projection, 263–66
    perspective projection, 267–72
Concurrent vectors, 164
Cone,
    by rotation, 113
    method for true length, 51
    development of, 285
Constant of integration, 233
Conventional symbols, 307–16
Convolute,
    geometry, 289
    development, 290
Coordinates,
    general, 37
    isometric, for computation, 257–61
    perspective, for computation, 267–71
Coplanar vectors, 164
    parallel, 171
Couple, resultant, 172
Curved surfaces, 283
Cutting planes,
    for intersections, 279
    inclined, 281
    for piercing points, 105
    symbol, 302
Cylinder,
    by rotation, 113
    development of, 284

Datum, 306
Depth, 16
Developments, 283
    cone, 285
    convolute, 289–90
    cylinder, 284
    elbow-joint, 286
    sphere, 291
    transition piece, 287
    warped surface, 288
Dihedral angle,
    by auxiliary plane, 73
    by rotation, 123

Dimensions, 305
   holes, 306
   lines, 302
   pictorial views, 142
Dimetric sketching, 345
Distance,
   perpendicular, from point to line, 89–92
   perpendicular, between two skew lines, 93–96
   perpendicular, from point to plane, 99
   shortest horizontal between two skew lines, 97
Dividers, 358
Double-curved surfaces, 283
Drawing board, 355
Drawings,
   dimetric, 345
   isometric, 137–41, 257–66
   oblique, 133, 159
   one-view, 300
   orthographic, 19, 23–29
   perspective, 143–58, 267–72
   pictorial sketching, 344–50
   standard views, 19, 23, 25, 27, 29
   trimetric, 345
   two-view, 301

Edge views,
   for intersections, 277
   of plane, 69
   for piercing point, 103
Elbow joint, development, 286
Electrical drawing symbols, 313
Ellipses, 141
Engineer's scale, 359
Equations, for computation, 261, 269
Equilibrium, 167
   coplanar vectors, 182
   noncoplanar vectors, 197
Equipment, 355–59
Exploded views, 349

Fits, 318
Folding lines, 39
Forces, *see* Vectors
Freehand technical sketching, 339–50
French curve, 12
Front view, 19, 23, 301
Frontal line, 43
Frontal plane, 19, 25, 27
Funicular polygon, 168

Gear, forces on helical, 174
Geometric solids, 2, 3
Geometrical tolerancing, 320–30
Gore development, 291
Graphical calculus,
   differentiation, 227–32
   integration, 233–35
Ground line, 143
Guide lines, for lettering, 366

Half section, 304
Height, 16
Hidden lines, 79, 302
Holes, dimensioning, 306
Horizon line, 143
Horizontal plane, 19
Hyperboloid, 113

Inked lines, 301
Instruments, 355
Integration, graphical, 233–35
Interference fits, 319
Interpreting engineering drawings, 299–330
Intersections,
   edge-view method, 277
   cutting-plane method, 279–81
Intersecting lines, 57
Isometric projection, 135–38
   by calculator or computer, 257
   circles, 141
Isometric drawing, 139–41
   freehand sketching, 344

Joints, method of, 179

Kinematics,
   displacement, 209
   position synthesis, 210–17
   space mechanism, displacement, 219
   space mechanism, velocity, 221

Lay, 307
   symbols for, 308
Lettering, 365–69
Limits and fits, 318
Lines in space, 39–42
   at specified angles, 119
   bearing, 49
   frontal line, 43

horizontal line, 43
intersecting, 57–60
non-intersecting, 61
perpendicular to a plane, 77
point view, 53
profile line, 43
slope of, 49
true length, 49, 51, 194
Linework, 301–02

Mechanisms, 209
planar, synthesis, 210
space, displacement, 219
space, velocity, 221
spatial, synthesis, 213–16
three-position synthesis, 217
Meridian development method, 291
Metric,
scales, 360
threads, 309

Nonconcurrent forces, 183
Noncoplanar vectors,
addition, 189
edge-view method, 202
equilibrium, 197
point-view method, 197
resultant, 191–96
trial-solution method, 201
Numerals, lettering, 369

Oblique projection, 129–34
cabinet, 133
cavalier, 133
from orthographic, 131
Oblique drawing, 159
freehand, 344
One-view drawings, 300
Origin,
for locating points, 37
for perspective by computation, 267
Orthographic projection, 15–22
first-angle, 23–26
third-angle, 27–30
Orthographic freehand sketching, 339

Paper, drawing, 355
Parallel forces, 171
Parallel lines, 65

Parallelepiped, 2
Parallelogram law, 163
Parts list, 317
Pencils, 355
Perpendicular lines,
intersecting, 57–60
non-intersecting, 61
perpendicular to a plane, 77
Perspective projection, 143
by calculator or computer, 267–72
freehand, 346–48
from orthographic views, 145–48
vanishing points, 149–56
Photogrammetry, short range
flat ground, 249
photographic view, 239
vertical objects, 239
Pictorial views,
by calculator or computer, 257–72
dimensioning, 142
dimetric, 345
exploded, sketching, 349
freehand sketching, 339–50
isometric, 135–41, 344
oblique, 129–34, 344
perspective, 143–58, 346
trimetric, 345
Piercing points,
edge-view method, 103
cutting-plane method, 105
Piping, 313
Planes, 67
angle between, 73, 123
auxiliary, 49, 53
cutting, 105, 279, 281, 302
distance to a point, 99
edge-view, 69
frontal, 19, 25, 27
horizontal, 19
inclined auxiliary, 53
intersection of, 277–82
line perpendicular to, 77
profile, 19
true shape, 71, 121
true-slope angle, 69
Plotting of points, see Coordinates
Plotting machines, 257
Point view of line, 53
method for vectors, 197

Points,
    location by coordinates, 37, 257, 267
    projection, 35
Position tolerances, dimensioning, 329, 330
Prism, 2
Profile line, 43
Profile plane, 19
Projection,
    auxiliary, 49, 53
    cabinet, 133
    cavalier, 133
    dimetric, 345
    isometric, 135, 141, 257, 344
    oblique, 129
    orthographic, 15–22
    perspective, 143, 267
    trimetric, 345
Projection planes,
    unfolding, 19
    first-angle, 23–26
    third-angle, 27–30
Protractor, 358

Rate of change, 227
Reaction forces, 173, 177
Rear view, 23, 27
Relative velocity, 217
Removed sections, 305
Resolution of a vector, 164
Resultants of vectors, 167, 172, 191
Revolved sections, 304
Rivets, 314
Rolled shapes, 314
Rotation,
    dihedral angle, 123
    line about a line, 113
    point about a line, 111
    position of line at specified angles, 119
    true length, 51
    true shape, 121
Roughness, surface, 307

Scales, 359
    architect's, 359
    engineer's, 359
    metric, 360
Screw threads, 309–10
Seam line, 284
Section lines, 302, 303
Sectional views, 303–05
Sections, method of, 184
Shading, 350
Shadows, 108
Shortest distance, 93, 97
Side view, 19, 23, 27, 301
Single curved surfaces, 283
Simplified drawings, 317
Sketching, freehand, 339–50
Skew lines, 93–98
Slope of a line, 49, 227, 229
Slope angle of a plane, 69
Space mechanism,
    displacement, 219
    velocity, 221
Space quadrants, 21
Sphere, development, 291, 292
Station point, 143
Steel structural shapes, 314
Stereoscopic pairs of drawings, i
Stretch-out line, 284
Structural drawings, 314
Surface finish, 307
Symbols for drafting,
    architectural, 316
    electrical and electronic, 313
    piping, 312
    rivets, 315
    screw threads, 309
    surface finish, 307–08
    welding, 311
Synthesis,
    plane mechanism, 210, 217
    space mechanism, 213–16

T-square, testing, 356
Testing, drawing instruments,
    T-square, 356
    parallel rule, 356
    triangles, 357
Tangent,
    to a circle, 10
    to curves, 227–32
Threads, types, symbols, 309
Tolerances, 318, 319
    geometrical, 320–30
Top view, 19, 23, 27, 301
Torus, 298
Transition fit, 319

Transition piece, development 287
Triangles, instrument, 357
Trimetric sketching, 345
True angle between line and oblique plane, 75
True length of a line,
    by auxiliary plane, 49
    by rotation, 51
    shorter method, 194
True position dimensioning, 330
True shape,
    by auxiliary plane, 71
    by rotation, 121
Truss analysis, 176
    method of joints, 179
    method of sections, 184
Two-view drawings, 301

Unified threads, 309

Vanishing points, 149–56
Vector geometry,
    addition, 163, 189
    Bow's notation, 165
    coplanar parallel forces, 171
    coplanar vectors, 164
    edge-view method, 202
    equilibrium, 167, 182, 197
    funicular polygon, 168
    method of joints, 179
    method of sections, 184
    noncoplanar, 189
    point-view method, 197
    resultants, 167, 172, 191
    resolution, 164
    trial solution, 201
Velocity, relative, 217, 221
Views,
    auxiliary, 49, 53
    edge, 69, 103, 277
    front, 19, 23, 27, 301
    orthographic, *see* Orthographic projection
    rear, 23, 27, 301
    sectional, 303, 305
    side, 19, 23, 27, 301
    top, 19, 23, 27, 301

Warped surface, 283
    development, 288
Welding, 310
Width, 16
Word spacing, 369
Words, lettering, 369

X-Y coordinates, 258, 261, 267
X-Y-Z coordinates,
    general location, 37
    isometric, computation, 257
    perspective, computation, 267